Practical Physical Chemistry

Practical Physical Chemistry

B. Viswanathan
P.S. Raghavan

Viva Books

New Delhi | Mumbai | Chennai | Kolkata | Bengaluru | Hyderabad | Kochi | Guwahati

First Published 2005
This edition published in 2012 by
Reprinted 2014, 2015, 2017, 2019, 2023, 2026

VIVA BOOKS PRIVATE LIMITED

- 4737/23 Ansari Road, Daryaganj, New Delhi 110 002
  E-mail: vivadelhi@vivagroupindia.net, Tel. 42242200
- 76, Service Industries, Shirvane, Sector 1, Nerul, Navi Mumbai 400 706
  E-mail: vivamumbai@vivagroupindia.net, Tel. 27721273, 27721274
- Megh Tower, Old No. 307, New No.165, Poonamallee High Road, Maduravoyal
  Chennai 600 095, E-mail: vivachennai@vivagroupindia.net, Tel. 23780991, 23780992
- B-103, Jindal Towers, 21/1A/3 Darga Road, Kolkata 700 017
  E-mail: vivakolkata@vivagroupindia.net, Tel. 22836381, 22816713
- 7, Sovereign Park Apartments, 56-58, K. R. Road, Basavanagudi, Bangalore 560 004
  E-mail: vivabangalore@vivagroupindia.net, Tel. 26607409, 26607410
- 101-102 Mughal Marc Apartments, 3-4-637 to 641, Narayanguda, Hyderabad 500 029
  E-mail: vivahyderabad@vivagroupindia.net, Tel. 27564481, 27564482
- First Floor, Beevi Towers, SRM Road, Kaloor, Kochi 682 018, Kerala
  E-mail: vivakochi@vivagroupindia.net, Tel: 0484-2403055, 2403056
- 232, GNB Road, Beside UCO Bank, Silpukhuri, Guwahati 781 003
  E-mail: vivaguwahati@vivagroupindia.net, Tel: 0361-2666386

www.vivagroupindia.com

ISBN : 978-81-309-2069-6

Published by Vinod Vasishtha for Viva Books Private Limited, 4737/23, Ansari Road, Daryaganj, New Delhi 110 002.

Printed and bound in India by Ashim Print Line, Sahibabad.

CONTENTS

PART XI: VOLTAMETRY

PART XII: MISCELLANEOUS EXPERIMENTS

PREFACE

The teaching and learning of Science essentially depend how well the laboratory curriculum is designed and practised. It is often said that our primary and secondary education are of very high standard compared to the corresponding education elsewhere in the world but our tertiary education somehow appears to lag behind other countries. One of the possible reasons can be that our curriculum does not have firm experimental base. Scientific developments depend on experimentation and also good practical training and in these spheres of activity in the tertiary education, India seems to be far behind compared to other countries. The laboratory curriculum of the undergraduate studies of some of the western universities have reached the stage which we cannot even match at our research level and this disparity is causing considerable imbalance at the scientific research output of our country. The authors have examined this aspect carefully for a number of Indian Universities and evolved a common minimum laboratory programme in Physical Chemistry and the result is the present compilation. The experiments chosen are the minimum required for our undergraduate programme and we have purposely included some experiments which can be covered at least at demonstration level and can be given as exercises at the post-graduate level.

The authors have attempted to assemble the list of experiments from their experience and also have drawn upon the experience of the students who have undergone these laboratory courses and felt the inadequacy of the existing curriculum. The authors are aware that they have not yet exhausted all the experiments that they wanted to place in this book but they had to make a selection keeping the size in consideration. Few years back one of the authors was involved in a syllabus revision for a few institutions and the exercises included on spectral interpretation was a result of this experience and the authors have drawn from the original booklet prepared by a team of people including one of the authors of this book and they are included in this book to a very limited extent only.

The authors have drawn from the experiences of many of their colleagues in their respective institutions and they wish to express their gratitude to each one of them though not by name. The excellent academic atmosphere and stimulating research environment one of the authors enjoyed at the Department of Chemistry, Indian Institute of Technology, Madras have also contributed to this venture and hence he is duty bound to acknowledge with gratitude the authorities of this institution for providing this atmosphere.

The authors have certain specific motivation in bringing out this publication and it will be fulfilled only if the stake holders namely the teachers, students and other users of this publication can bring to our attention the shortcomings and other issues to make this publication more beneficial to the community to which it is meant. The authors will be more than willing to take all comments and incorporate them in the further editions of this volume.

Finally it is our duty to thank many of the invisible people who have contributed in no small measure to this endeavour.

Authors

WORK ETHICS FOR A CHEMICAL LABORATORY

As any other endeavour in science, working in a chemical laboratory also requires following certain work ethics. We briefly mention them below:

1. Never work in a laboratory without proper supervision.
2. Use laboratory coat or apron while working in the laboratory.
3. Always handle the chemicals carefully.
4. Do not carry the reagent bottles to your work place. Take a suitable watch glass or a beaker to the reagent shelf and bring sufficient quantity of the reagent to your work table.
5. Do not weigh any substance on a piece of paper. Use proper weighing bottles. Keep the balance and its surroundings clean. Do not spill the chemicals on the pan of the balance.
6. If you have to insert a glass tube to rubber, make sure the glass tube is fire-polished, hold the glass near the rubber tube, cover your hand with a towel and do not force the glass tube into the rubber, gently use a twisting motion and if necessary use a lubricant, like water or glycerine.
7. Never taste any chemical in the laboratory.
8. Whenever you have to smell a chemical, use your hand as a fan to direct the odour in the direction of your nose and *never* place your head directly over the container.
9. If poisonous gases are evolved in your experiment, make sure that the experiment is carried out with proper exhaust system.
10. If you have to dilute an acid, always add acid to water with constant stirring.
11. Place all waste papers, used matches, broken glass and other material in the waste basket.
12. For soluble solid wastes, put them in the sink and dissolve them with a large quantity of water and drain them.
13. Acids, bases and other liquids must be drained down the sink with water.

14. All insoluble solid wastes (for example, many organic chemicals used in the phase rule studies) should be deposited in specially assigned containers.
15. Never pour back the reagents taken by you in the reagent bottles.
16. Do not directly insert a pipette into the stock reagent bottles.
17. Make sure that you have taken the correct reagent. Read the label of the reagent bottles and never go by any other marks and also label all your solutions in your workbench.
18. Reagent bottles should be properly closed (by replacing the cover) after reagent is withdrawn by you for use. Make sure that the reagent bottles are not left without cover.
19. Never place the cover or cork of the reagent bottle on the bench.
20. All glass tubes, rods, spatula should have been properly fire-polished. It may be necessary to know how to fire-polish a tube in a flame. Make yourself familiar for fire-polishing glass tubes and rods.
21. Graduated measuring cylinders or graduated bottles should not be heated.
22. Keep the observation books and papers away from the working place.
23. Before entering the laboratory make sure the location of the safety measures like fire extinguisher, emergency shower, first aid kit.
24. Before leaving the laboratory, make sure you clean your work place and other points of the laboratory where you have carried out some operation like transferring, weighing, etc.
25. Never spill mercury from the electrodes and in the case of spillage of reagents use sufficient quantity of water to wash the place immediately. In the case of other spillage use proper precautions to remove them.
26. Report all accidents, injuries, however minor they may be, to the instructor.
27. It is advisable that you familiarize yourself with necessary first-aid instructions for acid, alkali and other chemical burns.
28. When you have to use an equipment, make sure that you have read all the instructions before plugging the instrument into the mains.
29. While using gas cylinders make sure that you fix proper and appropriate regulators and the cylinders are kept securely tightened with chains.

PART I

Determination of Molecular Weight

EXPERIMENT 1
Rast's Method

Aim: To determine the molecular weight of the given substance by the Rast's method.

Apparatus Required: Boiling tube, stirrer, thermometer and stop clock.

Chemicals Required: Naphthalene and biphenyl.

Principle: When a liquid or a dilute solution is continuously cooled, a solid begins to separate and the temperature at which this solid separation begins is called its freezing point. Thus at the freezing point, the solution is in equilibrium with the solid. Generally, solutions freeze at lower temperature than that of the pure solvent. This freezing point lowering of a solution is a direct consequence of the vapour pressure lowering of the solvent by dissolved solute. The lowering of freezing point, ΔT_f is related to the amount of the solute in the solution as $\Delta T_f = K_f \times (w_2/M_2) \times (1000/w_1)$ where w_1 is the mass of solvent, w_2 the mass of the solute added, M_2 the molecular mass of the solute and K_f is the cryoscopic constant, which is constant for a given solvent. By taking a solute of known molecular weight and measuring the ΔT_f value, the K_f for the solvent is determined first; then the experiment is repeated with the given solute and its molecular weight is determined using the K_f value calculated earlier.

Procedure: Take about 5 g of naphthalene in a boiling tube. Melt it by heating; allow the melt to cool and note the temperature for every 30 seconds, till the temperature remains constant for at least two minutes. Plot the temperature against time. This is called the cooling curve. Find out the melting point from the cooling curve of the molten liquid. Then add 0.1 g of the given substance (biphenyl) to the boiling tube. Melt the mixture. Obtain the cooling curve for the solution and determine its freezing point. Repeat this procedure for the addition of 0.2 g and 0.3 g of solute and note the depression in freezing point values for each composition of the solute.

Treatment of Data: Tabulate the observed data (temperature) in the following format.

S.No.	Time (s)	Solvent(naphthalene) alone	With 0.1 g solute	With 0.2 g solute	With 0.3 g solute

Note the freezing point of the solvent and the solutions from the respective cooling curves.

In each case calculate the value of depression in freezing point as $\Delta T_f = T_0 - T$, where T_0 and T are the freezing points of the solvent and the solution respectively. Using the ΔT_f value, arrive at the cryoscopic constant for naphthalene from the formula

$$\Delta T_f = K_f \times (w_2/M_2) \times (1000/w_1)$$

Take the average of K_f. Now, using the average K_f value, calculate the molecular weight of the solute. Take the mean of the three calculated molecular weights.

Results:

(i) Cryoscopic constant for naphthalene = ... deg mol^{-1} per kg of solvent.

(ii) Molecular weight of the given solute = ...

Table 1: Values of the molar depression in freezing point constant or the cryoscopic constant for some common solvents.

Solvent	K_f deg mol^{-1} per kg of solvent
Acetic acid	3.90
Benzene	5.10
Bromoform	14.3
Camphor	40.0
Cyclohexane	20.2
Nephthalene	7.0
Nitrobenzene	6.90
Water	1.86

Table 2: Characteristics of probable solutes for molecular weight determination by cryoscopic method.

Solute	Melting point (°C)	Molecular weight
Diphenyl	69.7	154.2
p-Dichlorobenzene	53.0	147.0
o-Dichlorobenzene	17.0	147.0
p-Nitrotoluene	51.7	137.2

EXPERIMENT 2
Beckmann Method

Aim: To determine the molecular weight of a substance from the depression of freezing point—Beckmann method.

Appartus Required: Beckmann thermometer, glass stirrer, freezing point tube, thermometer and one litre beaker.

Chemicals Required: Glucose or urea, distilled water or benzene, ice and salt.

Principle: It is based on the fact that the freezing point of a solution of non-volatile solvent is lower than that of the pure solvent and that the depression in freezing point is proportional to the concentration of the solution. The lowering of freezing point, ΔT_f is related to the amount of the solute in the solution as $\Delta T_f = K_f \times (w_2/M_2) \times (1000/w_1)$ where w_1 is the mass of solvent, w_2 the mass of the solute added, M_2 the molecular mass of the solute and K_f is the cryoscopic constant, which is constant for a given solvent.

In order that the determination of the depression of the freezing point can be made with sufficient accuracy, it is necessary to read the temperatures with an error not exceeding 0.001 to 0.002°C. A Beckmann thermometer is therefore used (Fig.1). This thermometer has a large bulb and a range of only 5 or 6 degrees. It is graduated to 0.01°C and can be read to the nearest 0.001°C, with the aid of a special lens. It does not indicate absolute temperatures, but merely difference of temperature. The special feature of the Beckmann thermometer is that the amount of mercury in the bulb can be adjusted to make the thermometer read at any desired region of temperature. The regulation is effected by transferring some mercury into a small reservoir at the upper end of the capillary or vice versa.

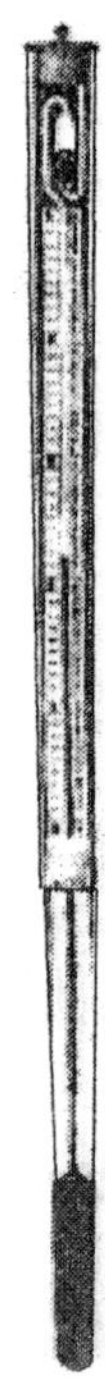

Fig. 1. Beckmann Thermometer

Procedure: The Beckmann thermometer can be set up by placing it in a bath maintained at 5°C for benzene or at 0°C for water. Pipette out exactly 25 cm^3 of the given solvent (benzene or water), into a freezing point tube. Place the Beckmann thermometer and the glass stirrer carefully into the freezing point tube. The bulb of the thermometer must be completely immersed in the solvent. Keep the entire freezing point tube in freezing bath (containing ice and a small amount of salt) maintained at 0°C. Stir the solvent vigorously with the help of the glass stirrer. The temperature decreases continuously and when the crystallization of the liquid commences the temperature rises up and attain a constant value. Note this constant value. Again melt the frozen solvent (by keeping it in a beaker of ordinary water at room temperature) and determine the freezing point of the solvent as above. Repeat the above procedure at least three times to get the correct freezing point of the solvent. It is convenient to have the mercury level at the higher graduation levels of the thermometer at the time of setting it for solvent. After setting it for solvent, add 0.2 g of a substance of known molecular weight and note the fall in temperature. Repeat this for one more addition of the same substance. Then pour the solution out and clean the freezing point tube. Pipette out 25 cm^3 of fresh solvent into it. Introduce 0.2 g of the substance whose molecular weight has to be determined, into the freezing point tube. Dissolve the substance well and determine the fall in temperature of the solution as described above. Add again one more 0.2 g of the substance and find out the fall in temperature.

Treatment of Data:

Density of the solvent = . . . g cm^{-3}

Cryoscopic constant for the solvent used = . . . deg mol^{-1} per 1 kg of solvent

Record your observations of the freezing point of the solvent and solutions in the following table; calculate the K_f value for the solvent and the molecular weight of the unknown substance using the formula

$$\Delta T_f = K_f \times (w_2/M_2) \times (1000/w_1)$$

System	Beckmann thermometer reading	ΔT_f (°C)
Solvent	T =	
Solvent + 0.2 g of known substance	T_1 =	$T - T_1$ =
Solvent + 0.4 g of known substance	T_2 =	$T - T_2$ =

System	Beckmann thermometer	ΔT_f
Solvent + 0.2 g of unknown substance	T_3 =	$T - T_3$ =
Solvent + 0.4 g of unknown substance	T_4 =	$T - T_4$ =

Results:

(i) Cryoscopic constant of the solvent, K_f= .... deg mol^{-1}/kg of solvent.

(ii) Molecular weight of the given substance = ...

EXPERIMENT 3
Viscosity Measurements

Aim: To determine the molecular weight of a given polymer from viscosity measurements.

Appartus Required: Viscometer, thermostat, pipettes and stop watch.

Chemicals Required: Benzene, a polymer material (rubber or any polymer soluble in benzene).

Principle: If a liquid with a coefficient of viscosity η flows with a uniform velocity at a rate of V cm^3 in t seconds through a narrow tube of radius r cm and length l cm under a driving pressure of p dyne/cm^2 then $\eta = \pi r^4 pt/8Vl$.

This equation was derived by Poiseulle and is the basis of methods for determining coefficient of viscosity of liquids as well as gases. The direct measurement of viscosity requires the determination of the rate of flow of liquid through a capillary tube of known dimensions and the application of the above equation. Viscosities of two liquids may be compared; if the coefficient of one of them is known, that of the other can be evaluated. Since the applied pressure is proportional to the density (ρ) of the liquid, it follows from the above equation—since the same capillary tube is employed i.e., r and l are constant and the same volume V of the liquid flows through it in each case, that for the two liquids 1 and 2, $\eta_1/\eta_2 = (\rho_1/t_1) \times (\rho_2/t_2)$, where t_1 and t_2 are the time of flow.

It was noted by H. Staudinger that the viscosity of a dilute solution of a polymer in a suitable solvent varies in a regular manner with the molecular weight of the polymer. The higher the molecular weight in a homologous series of linear polymers, the greater is the increase in viscosity for a given weight concentration of polymer. Originally derived empirically, the relationship between viscosity and molecular weight has been shown to have a theoretical basis.

The presence of large molecules in a liquid medium introduces inhomogeneities which increase the resistance to flow of the molecules of liquid and thus increase viscosity. If η_o and η are viscosities of the solvent and solution respectively, η/η_o is called the relative viscosity, η_r. Also the ratio, $(\eta - \eta_o)/\eta_o = \eta_r - 1$, known as the specific viscosity, η_{sp}. It is the ratio η_{sp}/C referred to as the relative increase in viscosity per unit concentration of polymer which is a function of the molecular weight of the polymer. Since this ratio varies with concentration, it is necessary to extrapolate a plot of η_{sp}/C against C to infinite dilution and the value thus obtained is called the intrinsic viscosity, $[\eta]$, so that $\underset{C \to 0}{\mathrm{Lt}} (\eta_{sp}/C) = [\eta]$.

Since η_{sp} is dimensionless, $[\eta]$ has the units of reciprocal concentration and since C is usually expressed in gram per L of solution, $[\eta]$ is given in deciliter per gram. The relationship between the intrinsic viscosity and the molecular weight of the dispersed polymer is given as $[\eta] = K M^{\alpha}$, where K is a constant for a given polymer—solvent system; M is the molecular weight of the polymer and α depends on the shape of the molecule. Theory predicts that α should be zero for large rigid spheres; 0.5 – 0.8 for randomly coiled (flexible) linear chains and 2 for rigid rods. The highest value found for α is 1.5 for amylase which has been shown by light scattering measurements to be rod shaped.

To obtain molecular weight from viscosity data, K and α in the above equation must be known. However, they can be determined only if molecular weights of samples of the given polymer material have been obtained by other ways, i.e., by osmotic pressure, sedimentation or light scattering measurements. A plot of $[\eta]$ against log M will give straight line of slope α and intercept log K. It is thus seen that the viscosity method is a secondary (or dependent) method for determining molecular weights in contrast to other methods which are absolute and yield M from directly measurable quantities. However, once K and α are known for a given type of polymer and solvent, the viscosity method becomes one of the most convenient for the subsequent evaluation of molecular weight. The value obtained is a number average molecular weight which differs slightly from the weight average molecular weight.

Procedure: Take accurately about 12 cm^3 of benzene in one of the arms of the viscometer. Place the viscometer in the thermostat maintained at 25°C. The viscometer has three arms. Attach a clean rubber tubing to the capillary end of the viscometer. Suck benzene such that the liquid meniscus stands above the indicated mark on the arm. Release the liquid from the above position and allow it to fall back. When the meniscus passes the indicated mark, start the stop watch. Note the time of flow between the two marks. Repeat the above procedure to get concordant values. Clean and dry the viscometer and repeat the experiment with 14 cm^3 of the polymer solution. Similarly determine the time of outflow for different concentrations of the polymer solution. This is done by adding 2 cm^3 of benzene to the polymer solution in the viscometer and times of outflow are determined for four such solutions.

Treatment of Data: Tabulate the observed time of flow for the systems as shown below.

Volume of polymer	Mean time of flow (s)	η_{rel}	η_{sp}	Concentration g/100 cm^3	Total volume	η_{sp}/C
	t_1					
	t_2					
	t_3					
	t_4					
	t_5					

In the table, t_1 is the time of flow for the solvent, benzene and t_2, t_3, etc. are the time of flow for the various concentrations of the polymer solution. Compute the relative viscosities as the ratio of t_2/t_1, t_3/t_1, etc. The reciprocal of each of these relative viscosities would give the specific viscosities, $\eta_{sp.}$ Dividing each of these values by concentration of the polymer solution, C yields η_{sp}/C. Plot a graph of η_{sp}/C on the y–axis and the concentration, C on the x-axis. A straight line results. It is extrapolated and the intercept gives the intrinsic viscosity, η_{int}. The molecular weight of the polymer is then calculated using the relation $M = KM^{\alpha}$.

Results:

(i) The intrinsic viscosity of the polymer solution = . . . deciliter/g

(ii) The molecular weight of the polymer = ...

EXPERIMENT 4
Steam Distillation

Aim: To determine the molecular weight of the higher boiling constituent of a mixture using the principle of steam distillation.

Appartus Required: Steam can, round bottom flask 1L capacity, still head, condenser, 100, 50 and 25 cm^3 measuring cylinders, thermometer, tripod stands, 2 iron stands, 2 clamps and 2 burners.

Chemicals Required: Nitrobenzene, distilled water.

Principle: In a mixture of two immiscible liquids, each will exert the vapor pressure corresponding to the liquid (pure) at the given temperature and the total vapor pressure will be the sum of the vapor pressures of the two pure materials, $P = P_A^o + P_B^o$ where P is the total pressure and P_A^o and P_B^o are the vapour pressures of the two pure liquids A and B.

If N_A and N_B be the mole fractions of the two constituents in the vapour, then.

$$P_A^o = N_A P \qquad \text{and} \qquad P_B^o = N_B P$$

Hence $$P_A^o / P_B^o = N_A P / N_B P = N_A / N_B$$

But $N_A = n_A/(n_A + n_B)$ and $N_B = n_B/(n_A + n_B)$ where n_A and n_B are the number of moles of A and B in any given volume of vapour. Consequently, $P_B^o / P_B^o = n_A/n_B$ must also be a constant, i.e., the composition of the vapour is constant as long as both the liquids are present. Further, $n_A = W_A/M_A$ and $n_B = W_B/M_B$ where W_A and W_B are the weights in any given volume of the vapour and M_A and M_B are the molecular weights of A and B respectively, the above equation becomes

$$\left(P_A^o / P_B^o\right) = n_A/n_B = (W_A/W_B)\,(M_A/M_B)$$

and $$W_A/W_B = M_A P_A^o / M_B P_B^o$$

The ratio of the weights of the constituents in the vapour is the same as the ratio in the distillate.

Distillation of immiscible liquids is utilized industrially and in the laboratory for the purification of organic liquids which either boil at high temperatures or tend to decompose when heated to their normal boiling point.

Precedure: Set up the steam distillation unit as shown in Fig. 2. Place some nitrobenzene in the flask together with a little quantity of water. Heat the can first. When steam starts coming out of the nozzle, introduce the nozzle into the flask and start heating the flask simultaneously. Note down the barometer pressure. When the mixture starts boiling, allow the distillate to come continuously and note the boiling point of the mixture on the thermometer, whose bulb is in the distilling vapour when the reading in the thermometer is steady. Collect the samples in graduated cylinders. Allow the distillate in the measuring cylinder to separate into two distinct layers. Note the volume of both the layers.

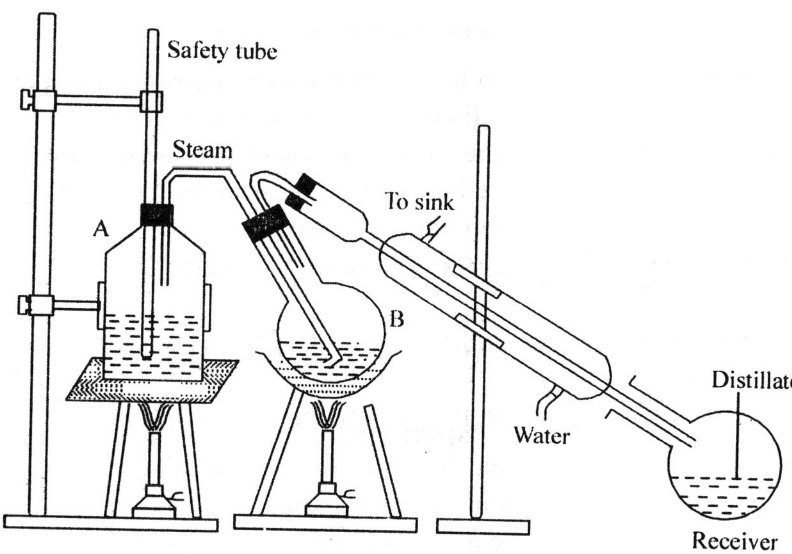

Fig. 2. Steam distillation unit

Treatment of Data:

Atmospheric pressure = . . . mm; Boiling point of the mixture = . . . ^{0}C

Vapour pressure of water at the boiling point of the mixture = . . . mm

Vapour pressure of organic liquid at the boiling point of the mixture = . . . mm

Density of the organic liquid = . . . g/cm^3

Volume of water (cm^3)	Weight of water (g)	Volume of organic liquid (cm^3)	Weight of organic liquid (g)

M_{org} = molecular weight of the organic liquid

W_{org} = weight of the organic liquid in g

W_{water} = weight of water in g

P_{water} = vapour pressure of water

P_{org} = vapour pressure of organic liquid at the boiling point of the mixture

M_{water} = molecular weight of water = 18.02g.

Calculate in each case in the above table, the molecular weight of the organic component using the equation

$$M_{org} = (W_{org}/W_{water}) \times (P_{water}/P_{org}) \times M_{water}$$

Result:

Report the molecular weight of the given organic substance.

PART II

Thermodynamic Properties of Solution

EXPERIMENT 5
Solubility of Naphthalene in Benzene

Aim: To study the solubility of naphthalene in benzene

Appartus Required: Burette, boiling tube, beaker, thermometer, and glass stirrer

Chemicals Required: Naphthalene and benzene

Principle: Solubility refers to the amount of a solute present in a saturated solution of a given solvent at a given temperature. It is an equilibrium phenomenon where the dissolved solute molecules are in equilibrium with both the solvent and the undissolved solid solute.

In the experimental study of both solubility and freezing points of solutions, what is determined is a temperature and the composition of the solution in equilibrium with excess of solid at that temperature. Whether the data are interpreted as solubility or freezing point, measurement depends upon the purpose of the experiment,

In any event, the equation $\ln x = L_f/R\ [(T_o - T)/T_oT]$

gives the variation of the freezing point (T) of an ideal solution with composition (mole fraction of solvent). Here, x is the mole fraction of solute, L_f is the molar heat of fusion and T_0 is the normal freezing point of the solute.

Procedure: Weigh 15 g of naphthalene into the boiling tube and place a thermometer with a stirrer. Add 6 cm^3 of benzene from the burette. Immerse the boiling tube in a beaker of warm water and stir until all the solid dissolves. Remove the tube from the warm water and allow it to cool in air. Note the temperature at which the first crystals separate. Add further 1 cm^3 portion of benzene and repeat the procedure to give a new (lower) crystallization temperature. Repeat the addition of 1 cm^3 portions of benzene and note the crystallization temperature every time.

Treatment of Data:

1. The ideal solubility is calculated using the equation

$$\ln x = [L_f/R]\ [(1/T_o) - (1/T)]$$

where x is the mole fraction of naphthalene in saturated solution; L_f is the molar latent heat of fusion of naphthalene = 4556.8 cal/mol. T_o is the melting point of pure naphthalene = 353 K and T is the crystallization temperature (K) and R is the gas constant.

2. To compare experimental and theoretical results, a graph of log x against 1/T is plotted using the experimental points. For this purpose the density of benzene is taken as 0.877 g/cm^3. Theoretical curve is drawn as line of slope $-L_f/2.303 \times R$ on the same axis through the point corresponding to the melting point of pure naphthalene (x = 1.0, T = 353 K) the other points in the line having as their ordinates, the ideal solubilities at various temperatures calculated from the above equation. (The slope of the theoretical line is found to be -1×10^{-3}).

Observations:

Weight of naphthalene (mg)	Volume of benzene in (cms)	Mole fraction of naphthalene	Crystallization temperature (°C)	Crystallization temperature (K)
15	6	0.6346		
15	7	0.5982		
15	8	0.5658		
15	9	0.5366		
15	10	0.5130		
15	11	0.4865		
15	12	0.4645		
15	13	0.4450		
15	14	0.4247		
15	15	0.4100		
15	16	0.3945		
15	17	0.3801		
15	18	0.3667		
15	19	0.3542		

Result:

The line obtained from experimental data should lie close to the theoretical line showing that the solution of naphthalene in benzene is almost ideal. Report the value of the latent heat of fusion of naphthalene from the experimental data.

Solubility of Naphthalene in Different Solvents

Solvent	Solubility*
Benzene	0.292
Toluene	0.289
CS_2	0.290
Nitrobenzene	0.295
CCl_4	0.257
Chlorobenzene	0.313
Hexane	0.121
Ethanol	0.024

* mole fraction of naphthalene

EXPERIMENT 6

Partition Coefficient for the Distribution of Iodine between Water and Carbon tetrachloride

Aim: To determine the partition coefficient for the distribution of iodine between water and Carbon tetrachloride.

Apparatus Required: Stoppered dry reagent bottles (250 cm^3), burette, pipettes, conical flasks, measuring cylinder and porcelain trough.

Chemicals Required: Potassium dichromate, sodium thiosulphate (0.1 N and 0.01 N) Potassium iodide, starch indicator solution, saturated iodine solution in CCl_4.

Principle: To a system of two liquid layers made up of two immiscible or partially miscible components, if a third substance soluble in both the layers is added it distributes itself between the two layers in a definite proportion. Such systems follow the 'Nernst distribution law' according to which the ratio of the concentrations (c) of the solute distributing itself between two solvents A and B is a constant (i.e) C_A/C_B = constant. Ususally, one of the liquids being organic and the other, water. The ratio is written as C_{org}/C_{aq} = K, the constant K is called the distribution coefficient or the partition coefficient. This is a direct consequence of the thermodynamic requirements for equilibrium. If the solute undergoes any chemical change in any one of the phases, the above law is applicable to the concentration of any particular species present in both the layers and not to the total concentration of the substance.

Procedure: Prepare 0.1 N $K_2Cr_2O_7$ solution by weighing accurately 0.490 g and dissolving it in 100 cm^3 distilled water in a volumetric flask. Using this solution, standardize the approximately 0.1 N solution and find out its accurate concentration. Then, dilute 25 cm^3 of this solution ten times of its volume accurately with distilled water. Label the two solutions with their accurate concentrations—one in the neighbourhood of 0.1 N and the other in 0.01 N.

Measure out (use burette or pipette out carefully) 20 cm^3 of the saturated iodine solution in CCl_4 into a well cleaned dry stoppered bottle. Add 200 cm^3 of distilled water to the above iodine solution. Place the mixture in an agitator for about 20 minutes; wait for sometime so that the organic and aqueous layers clearly separate. Then pipette out 25 cm^3 of the aqueous layer in a clean conical flask and titrate against the diluted sodium thiosulphate solution (0.01 N) taken in the burette, using starch indicator. Then pipette out 5 cm^3 of the CCl_4 layer into a clean conical flask; add 2 cm^3 of 10% KI solution and titrate against the 0.1 N sodium thiosulphate solution using starch indicator.

Treatment of Data:

Mass of $K_2Cr_2O_7$ = ... g

∴ Strength of $K_2Cr_2O_7$ solution = ... N

I. Standardisation of Sodium thiosulphate solution

$K_2Cr_2O_7$ – thio			Indicator: Starch
Volume of $K_2Cr_2O_7$ Solution (cm^3)	Burette reading (cm^3)		Volume of sodium thiosulphate (cm^3)
	Initial	Final	
20			
20			

∴ Concordant value = y cm^3

∴ Strength of sodium thiosulphate solution = $20 \times S/y$ N

II. Calculation of Partition Coefficient

Volume of aqueous layer pipetted out = 25 cm^3

Volume of ....N sodium thiosulphate solution consumed = cm^3

∴ $[I_2]$ in aqueous layer = ... N

= ... M

(Note: one mole of I_2 is twice its equivalent weight).

Volume of CCl_4 layer pipetted out = 5 cm^3

Volume of.........N sodium thiosulphate solution required = ... cm^3

∴ $[I_2]$ in CCl_4 layer = ... N

= ... M

∴ Partition coefficient (K) = $[I_2]\ CCl_{4\ \text{layer}} / [I_2]_{\text{aq. layer}}$ = ...

Result:

Partition coefficient of iodine between water and CCl_4 is...

EXPERIMENT 7

Equilibrium Constant for Homogeneous Equilibria and Determining the Concentration of a Given Solution

Aim: To determine the equilibrium constant for the system $I_2 + KI \rightleftharpoons KI_3$ and hence to determine the concentration of the given KI solution.

Apparatus Required: Stoppered dry reagent bottles (250 cm^3), burette, pipettes, conical flasks, measuring cylinder and porcelain trough.

Chemicals Required: Solid potassium dichromate, Sodium thiosulphate (0.1 N and 0.01N), Potassium iodide, starch indicator solution and saturated iodine solution in CCl_4.

Principle: Iodine is distributed between aqueous KI solution and CCl_4. The iodine present in the CCl_4 layer will be the molecular I_2 while that in the aqueous layer would react with KI, establishing the equilibrium, $I_2 + KI \rightleftharpoons KI_3$.

$\therefore$ The equilibrium constant, $K_{eq} = [KI_3]/[I_2]_{aq}\,[KI]$

The $[I_2]$ in CCl_4 is found out by titration with sodium thiosulphate solution; since the ratio, $[I_2]$ in $CCl_4/[I_2]_{aq}$ = Partition coefficient = 80, $[I_2]_{aq}$ can be found out. The concentration of combined iodine (in the form of complex, $[KI_3]$ can be calculated as the difference between the total $[I_2]_{aq}$ and $[I_2]_{free}$ in aqueous layer. From these data, the K_{eq} can be calculated. Then using the given test solution of KI, by redoing the experiment and using the K_{eq} value, the concentration of the given solution can be calculated.

Procedure: Prepare a 0.1 N solution of KI by weighing 4.15 g of the solid accurately and making it to 250 cm^3 in a volumetric flask. Measure out 15 cm^3 of saturated iodine solution in CCl_4 into a well cleaned dry stoppered bottle. Measure out 100 cm^3 of 0.1 N KI solution into the above bottle. Place the mixture in an agitator for nearly an hour and then keep in a porcelein trough containing water. After 20 minutes, pipette out 10 cm^3 of the KI layer and titrate against the 0.1 N sodium thiosulphate (which has been standardised) solution, using starch as indicator. Then pipette out 5 cm^3 of the CCl_4 layer in a conical flask, add 2 cm^3 of 10% KI solution and titrate against 0.1 N sodium thiosulphate solution, using starch indicator.

The procedure given above is then repeated with the given KI solution. Knowing the value of the equilibrium constant, one can calculate the strength of the given KI solution.

Treatment of Data:

I. Standardisation of sodium thiosulphate solution:

Mass of $K_2Cr_2O_7$ = W_1 g

Strength of $K_2Cr_2O_7$ solution made upto 100 cm^3 = $W_1 \times 10/49$ N

$K_2Cr_2O_7$–thio Indicator: Starch

S. No.	Volume of $K_2Cr_2O_7$ (cm^3)	Burette reading (cm^3)		Volume of thio (cm^3)
		Initial	Final	
1				
2				
3				

Concordant value = ...cm^3

Determination of Equilibrium Constant

Indicator: Starch

Vol. of 0.1N KI layer (cm^3)	Vol. of thio. solution (cm^3)	Normality (N) of iodine in KI layer	Vol. of CCl_4 layer (cm^3)	Vol. of thio solution (cm^3)	Normality of iodine in CCl_4 layer
10			5		
10			5		

Concentration of I_2 in CCl_4 layer = ... M

(Remember 2 equivalent of I_2 = 1 mole)

Total $[I_2]$ in KI layer = ... M. (i.e) $[I_2]_{aq} + [I_3^-]$

$\therefore$ $[I_2]$ combined in the form of I_3^- = $[I_2]_{total}$ – free $[I_2]_{aq}$

But, free $[I]_{aq} = [I_2]$ in CCl_4/partition coefficient

$= [I_2]CCl_4/80$

$\therefore$ $[I_2]$ combined $= \ldots$ mole/litre

Free $[I_3^-]$ at equilibrium $= [I_2^-]_{initial} - [I_3^-]$ combined

$= \ldots$ mole/litre

$\therefore$ Equilibrium constant, K $= [I_3^-]/[I_2][I^-] = \ldots$

Determination of the Strength of the given KI Solution

Indicator: Starch

Vol. of KI layer (cm^3)	Vol. of thio. Solution (cm^3)	Normality of iodine in KI layer	Vol. of CCl_4 layer (cm^3)	Vol. of thio solution (cm^3)	Normality of iodine in CCl_4 layer
10			5		
10			5		

$[I_2]$ in CCl_4 layer $=$ M

Free $[I_2]$ in aqueous layer $= [I_2]$ in CCl_4/partition coefficient

$= [I_2]\ CCl_4/80 = \ldots$ M

$[I_2]$ combined $= [I_3^-]_{aq} = [I_2]_{total} -$ free $[I_2]_{aq}$

$= \ldots$ M

Free $[I^-]_{equilibrium}$ $= [I^-]_{initial} - [I_3^-]_{aq}$

$= \ldots$ M

Since, the equilibrium constant, K $= [I_3^-]/[I_2][I^-]$

$[I^-] = [I_3^-]/\{[I_2]\ K\}$

$\therefore$ Free $[I^-]$ at equilibrium $= \ldots$ M

$\therefore$ Total concentration of the given KI solutions

$= [I_3^-]_{aq}$ + free $[I^-]$ at equilibrium

$= \ldots$ M

Results:

1. The equilibrium constant value for the $KI + I_2 \rightleftharpoons KI_3$ system is ....
2. The concentration of the given KI solution $= \ldots$ M

EXPERIMENT 8
Association Factor of Benzoic Acid

Aim: To determine the association factor of benzoic acid in benzene by distribution method (that is the distribution of benzoic acid between water and benzene).

Apparatus Required: Dry, well cleaned stoppered reagent bottles (150 cm^3); burette; 50, 20, 10 and 5 cm^3 pipettes; conical flasks; porcelain trough.

Chemicals Required: Benzoic acid, benzene, standard sodium hydroxide solutions 0.1 N and 0.02 N, phenolphthalein indicator.

Principle: When a substance is dissolved in two immiscible or partially miscible solvents, the dissolved substance, irrespective of its total amount, distributes itself between the two layers in a constant concentration ratio at constant temperature. This is known as the distribution law (also see experiment 7). When the dissolved substance does not have the same molecular weight in the two liquid phases, the simple form of the distribution law breaks down. The reason for this is that the law is applicable only to the molecular species that is common to both the layers. When benzoic acid is distributed between benzene and water, in water the acid consists mainly of simple benzoic acid molecules; there is some ionization, but this is small and may be ignored. In the benzene solution, on the other hand, the benzoic acid is largely associated to form dimers $(C_6H_5COOH)_2$, with a relatively small proportion of simple molecule, C_6H_5COOH. The distribution law holds for the species present in both layers, namely, the single C_6H_5COOH molecules. If allowance is made for the modification, the law is found to be obeyed on a satisfactory manner.

Suppose a given solute A has a normal molecular weight in one solvent in which the concentration is C_1, while in the second solvent, it exists mainly as the associated molecule A_n the total concentration being C_2. In the latter solvent there will be an equilibrium between the associated and simple molecules, i.e., $A_n \rightleftharpoons nA$.

And assuming the solution to be dilute, the equilibrium constant is, $K_c = C_A{}^n/C_{An}$ where C_A is the concentration of simple molecules and C_{An} of associated molecules in the second solvent. It then follows from the above equation that C_A = constant × $\sqrt[n]{C_{AN}}$.

If the solute is mainly in the associated form, C_{An} may be identified with C_2, the total concentration in the second layer; hence C_A = constant × $n\sqrt{C_2}$.

As stated above, the distribution law is applicable to the species that the two liquid phases have in common, namely, the simple A molecules; in this case. Therefore, it should take the form, C_1/C_A = constant, and upon introducing the value of C_A given by the above equation, the result is $C_1/\sqrt[n]{C_2}$ = constant.

The modified distribution law derived in this manner may be tested by means of the data for the distribution of benzoic acid between water and benzene.

Procedure: Add about 1 g of benzoic acid to a well stoppered bottle. Pipette out 100 cm^3 of water and 50 cm^3 of benzene into the same bottle. Shake the contents intermittently for 20 to 30 minutes and place it in the thermostat for the attainment of equilibrium. Pipette out 5 cm^3 of benzene layer into a conical flask containing about 10 cm^3 of water. Titrate this against 0.1 N sodium hydroxide solution using phenolphthalein as the indicator. Pipette out 20 cm^3 of the water layer and titrate against standard alkali (0.02 N). Repeat the above procedure by taking 2 g of benzoic acid.

Treatment of Data: Determine the concentrations C_1 and C_2 of benzoic acid in water and benzene layers respectively. Find out the values of C_1/C_2 and $C_1/\sqrt[n]{C_2}$.

C_1, M	C_2, M	C_1/C_2	$C_1/\sqrt{C_2}$

Observations:

Bottle No. 1

Volume of benzene layer, (cm^3)	Volume of NaOH (cm^3)	Volume of aqueous layer, (cm^3)	Volume of NaOH (cm^3)
5		20	
5		20	

Concentration of benzoic acid in aqueous layer C_1 = . . . M
Concentration of benzoic acid in benzene layer, C_2 = . . . moles/2

$$C_1/C_2 = C_1/\sqrt{C_2}$$

Repeat the calculations for the other bottles also.

Result:

The association factor of benzoic acid in benzene/water system is ———.

EXPERIMENT 9

Molecular Formula of Copper-ammonia Complex by the Partition Coefficient Method

Aim: To determine the partition coefficient of ammonia between water and chloroform and also to determine the formula of copper-ammonia complex.

Apparatus Required: Burette; 50, 25 and 20 cm^3 pipettes; stoppered bottles (2), 250 cm^3 conical flasks (2), wash bottle.

Chemicals Required: Approximately 1.0 N aqueous ammonia, 1.0 N and 0.05 N sulphuric acid, chloroform and methyl red indicator.

Principle: When an aqueous solution of ammonia is shaken up with chloroform, a particular concentration of ammonia gets distributed in the chloroform layer. At equilibrium, the ratio of the concentration of ammonia in the aqueous layer to that in the chloroform layer is a constant, known as the partition coefficient of ammonia between water and chloroform. Ammonia being weak base, its concentration in both the layers can be determined by titrating against standard sulphuric acid. This distribution law follows from phase rule and Henry's law and is a direct consequence of the thermodynamic requirements of equilibrium.

When Cu^{2+} solution is added to the ammonia–chloroform system, the ammonia molecule with its lone pair of electrons behaves as ligand and forms coordination complexes with copper (II) ion which is a d^9 system.

$$Cu^{2+} + xNH_3 \rightleftharpoons Cu(NH_3)_x^{2+}$$

The concentration of ammonia in the chloroform layer can be determined by titration against standard acid. The concentration of free ammonia in the aqueous layer can be determined, knowing the partition coefficient and concentration of ammonia in chloroform layer. Now, the total initial concentration of ammonia is equal to the sum of the concentration of ammonia in the chloroform layer, concentration of free ammonia in the aqueous layer and number of moles of ammonia taken up to form the complex.

So from all the other quantities, the number of moles of ammonia required to form the complex $Cu(NH_3)_x^{2+}$ can be determined.

Procedure:

I. Partition coefficient:

Pipette out 50 cm^3 of approximate 1 N aqueous ammonia solution and 50 cm^3 of chloroform into a stopperd bottle. Place the bottle in an agitator for some time and then keep it in a thermostat for about 15 minutes for equilibration.

Standardise the approximately 1.0 N sulphuric acid by titration of the acid taken in the burette against NaOH solution which has been already standardized using phenolphthalein as indicator. Dilute the 1.0 N sulphuric acid 20 times to get a 0.05 N sulphuric acid.

Analyse the chloroform layer for ammonia by titration of 20 cm^3 portions of it with 0.05 N sulphuric acid using methyl red as indicator. To aid the titration, add 20 cm^3 of water after 20 cm^3 of chloroform is pipetted out into a conical flask to extract the ammonia into the aqueous layer and then observe the colour change from yellow to red at the end point, in the aqueous layer. It would be necessary to add a drop of methyl red indicator towards the end point, for the detection of colour change. Otherwise, the end point cannot be found for all the methyl red would have been absorbed in the chloroform layer.

Similarly, estimate the concentration of ammonia in the aqueous layer by titrating 20 cm^3 portions of the aqueous layer with 1.0 N sulphuric acid solution using the same indicator.

II. Determination of formula of the copper-ammonia complex:

Pipette out 25 cm^3 of 1.0 N ammonia, 25 cm^3 of copper sulphate solution (0.1 M) and 75 cm^3 chloroform into a stoppered bottle. Prepare 0.1 M copper sulphate solution by weighing 2.5 g of it accurately and making upto 100 cm^3. Place the bottle in an agitator for some time and then put it in a thermostat for 15 minutes for equilibrium. Shake the bottle occasionally during equilibration. Standardise the approximately 1.0 N ammonia stock by titrating with 1.0 N sulphuric acid using methyl red indicator. Carefully withdraw 25 cm^3 portion of chloroform layer, add about 20 cm^3 of water and titrate against 0.05 N sulphuric acid using methyl red as indicator taking the precautions given earlier.

Observations:

Weight of $CuSO_4$ crystals = g

Normality of $CuSO_4$ solution = N

1. Determination of partition coefficient:

Volume of aqueous layer (cm^3)	Volume of 1 N sulphuric acid (cm^3)	Volume of chloroform layer (cm^3)	Volume of 0.05 N shulphuric acid (cm^3)
20		20	
20		20	

2. Determination of the formula of the complex:

Sulphuric acid versus NH_3 (N)

Volume of ammonia solution (cm^3)	Burette reading, (cm^3)		Volume of sulphuric acid solution, (cm^3)
	Initial	Final	
20			
20			
20			

3. Chloroform layer versus N sulphuric acid:

Volume of chloroform layer, (cm^3)	Burette reading, (cm^3)		Volume of sulphuric acid solution, (cm^3)
	Initial	Final	

Treatment of Data:

1. Determination of partition coefficient of ammonia between water and chloroform:

Volume of N sulphuric acid required for 20 cm^3 of aqueous layer = cm^3

Normality of ammonia in aqueous layer = . . . N

Volume of N sulphuric acid required for 20 cm^3 of chloroform layer = cm^3

Normality of ammonia in chloroform layer = . . . N

Partition coefficient = $[NH_3]aq/[NH_3]\ CCl_4$ = . . .

2. Determination of the formula of copper – ammonia complex:

Volume of N sulphuric acid required to neutralize 20 cm^3 of the ammonia solution = cm^3

Normality of the original ammonia solution, N_0 =

Normality of ammonia in chloroform layer, N =

Partition coefficient of ammonia between water and chloroform =

Normality of free ammonia in aqueous layer, N =

Moles of ammonia originally pipetted out into the bottle =

Moles of free ammonia in the aqueous layer at equilibrium =

Moles of free ammonia in chloroform layer =

Gram ions of Cu^{2+} pipetted out into the bottle =

Normality of $CuSO_4$ solution taken =

Moles of ammonia taken up to form the complex $Cu(NH_3)_x^{2+}$ =

Moles of ammonia originally pipetted out into the bottle =

Moles of free ammonia in the aqueous layer + moles of free ammonia in the chloroform layer + moles of ammonia taken up to form the complex $Cu(NH_3)_x^{2+}$.

Results:

1. Partition coefficient of ammonia between water and chloroform = . . .
2. Formula of the copper-ammonia complex $[Cu(NH_3)_x]^{2+}$ is ....

PART III

Thermochemical Experiments

EXPERIMENT 10
Heat of Solution of Oxalic Acid from Solubility Measurements

Aim: To determine the heat of solution of oxalic acid from solubility measurements.

Apparatus required: Burette, 1 cm^3 pipette, weighing bottle, conical flasks, thermometer, beakers.

Chemicals required: Sodium hydroxide, oxalic acid crystals.

Principle: In general solubility of substances increases with temperature. There are a few exceptions to this observation. In all cases, the effect of temperature on the solubility of a substance can be predicted from its heat of solution, provided no chemical reaction occurs during dissolution.

Le Chatlier's principle indicates that the substances which absorb heat during dissolution process (ΔH is positive), dissolve to a greater extent at higher temperature. In the same way it can be shown that the substances, which evolve heat (ΔH is negative) during dissolution dissolve to a greater extent at lower temperatures.

From thermodynamics, the effect of temperature on solubility is given by the relationship,

$$\log (s_1/s_2) = (\Delta H/2.303\ R)(1/T_2 - 1/T_1)$$

where s_1 and s_2 are the solubilities of the substances in η/100 g of water at temperatures T_1 and T_2, and R is the gas constant in cal deg^{-1} mol^{-1} and $T_2 > T_1$.

Procedure: Prepare saturated solutions of oxalic acid around 0°C, room temperature and 45°C. Keep them immersed in ice water maintained at 0°C, water at room temperature and in a thermostat at 45°C respectively. Note the room temperature and the temperature of the thermostat.

Quickly pipette out 1 cm^3 of the saturated solution at 0°C into a stoppered weighing bottle, previously weighed. Allow some time for the solution to attain room temperature. Wipe out the outside of the bottle by means of filter paper gently, to remove the condensed moisture and weigh again. From these two weights, the weight of the saturated solution around 0°C can be calculated.

Pipette out 1 cm^3 of the saturated solution of oxalic acid around 0°C into a conical flask. Titrate the solution against standard sodium hydroxide solution using phenolphthalein as indicator. Repeat the above procedure for saturated solutions at the other two temperatures.

Treatment of Data: From the normality of oxalic acid solution the weight of oxalic acid in 1 cm^3 of the solution can be calculated. Let it be x. If x_1 is the weight of 1 cm^3 of the saturated solution at 0°C, the solubility s_1 at 0°C is given by $s_1 = 100x/(x_1 - x)$.

Similarly, the solubilities s_2 and s_3 at room temperature, T_2 K and at the thermostat temperature, T_3 K can be calculated. The heat of solution of oxalic acid can be calculated from the equation.

$$\text{Log}\ (s_A/s_B) = [\Delta H/2.303R]\ \{(T_A - T_B)/T_A\ T_B\}$$

where s_A is the solubility at the lower temperature T_A K and s_B is the solubility at the higher tempeature, T_B.

Observations:

Normality of sodium hydroxide solution = . . . N

Normality of oxalic acid solution = . . . N

1. Temperature 0°C = 273.2 K

   Weight of weighing bottle + saturated solution = . . . g

   Weight of the weighing bottle = . . . g

   Weight of the saturated solution = . . . g

Volume of saturated solution (cm^3)	Burette reading (cm^3)		Volume of sodium hydroxide (cm^3)
	Initial	Final	
1			
1			

Normality of oxalic acid = . . . N

Weight of 1 cm^3 of oxalic acid solution = . . . g

Solubility of oxalic acid at °C $= s_1$

Repeat the calculation for T_2 and T_3

T_A	T_B	S_A	S_B	ΔH J/mol

Average heat of solution = ... J/mol.

T_A and T_B are any two of the three temperatures at which measurements have been made and s_A and s_B are the respective solubilities.

Result:

Heat of solution of oxalic acid = . . . J.

EXPERIMENT 11
Heat of Neutralisation of a Strong Acid by a Strong Base

Aim: To determine the heat of neutralization of a strong acid by a strong base.

Apparatus Required: Dewar flask with a glass hand stirrer, two 500 cm^3 standard flasks, thermometer (0 – 50°C).

Chemicals Required: 2.5 N Sodium hydroxide, 0.25 N Hydrochloric acid.

Principle: When a strong acid is neutralized with a strong base, a large quantity of heat is evolved; (i.e.) these neutralization reactions are exothermic in nature. As strong electrolytes, viz. HCl and NaOH remain as dissociated ions in solutions, the net reaction is,

$$H^+Cl^- + Na^+ OH^- \longrightarrow Na^+ Cl^- + H_2O.$$

Or, if it is between HNO_3 and KOH,

$$H^+ NO_3^- + K^+ OH^- \longrightarrow K^+ NO_3^- + H_2O.$$

Thus, irrespective of the identity of the acid or base, the net reaction is the same, viz: formation of water from H^+ and OH^-. Therefore, the heat evolved during these reaction is also a constant. This heat energy evolved is called the heat of neutralization and is being determined using the principles of calorimetry.

Procedure: The experiment consists of two parts.

1. Determination of water equivalent of the calorimeter:

Pipette out 50 cm^3 of distilled water into the Dewar flask. Allow sufficient time for thermal equilibration. Record the temperature. In a beaker, heat a quantity of water to about 50 – 60° C. Measure out 50 cm^3 of the warm water into another beaker and record its temperature. Add the warm water quickly in one bulk to the Dewar flask and stir well; record the final temperature attained. Calculate the water equivalent of the calorimeter.

2. Determination of heat of neutralization:

Pipette out 50 cm^3 of 2.5 N NaOH solution into the Dewar flask. Note the temperature of the solution. Take 250 cm^3 of 0.25 N HCl in a beaker. Record the temperature of the acid. Quickly pour the HCl into the Dewar flask containing NaOH . Start a stop watch and record the temperature every 15 seconds. Stir the mixture continuously and note the readings till a constant temperature is reached.

Treatment of Data

1. Water equivalent of calorimeter:

Temperature of the cold water = t_1°C

Volume of cold water taken in Dewar flask = 50 cm^3

Temperature of the warm water (before mixing) = t_2°C

Volume of warm water added = 50 cm^3

Temperature attained after mixing = t_3°C

Applying the principle, Heat gained = Heat lost

Heat taken up by the Dewar flask + water = $(M_1s + 50)(t_3 - t_1)$

Heat given out by warm water = $50(t_2 - t_3)$

∴ Water equivalent = $M_1s = \{[50(t_2 - t_3) - 50(t_3 - t_1)]/(t_3 - t_1)\}$

2. ΔH of neutralization**:**

Temperature variation on neutralization

Time (s)	15	30	45	60	75	90	105	120
Temp. (°C)								

Draw a graph of temperature-time. The temperature increases steeply on mixing, reaching a maximum and slowly decreases due to radiation losses. The differences between the maximum (a) and minimum (b) corresponds to $(t_5 - t_4)$.

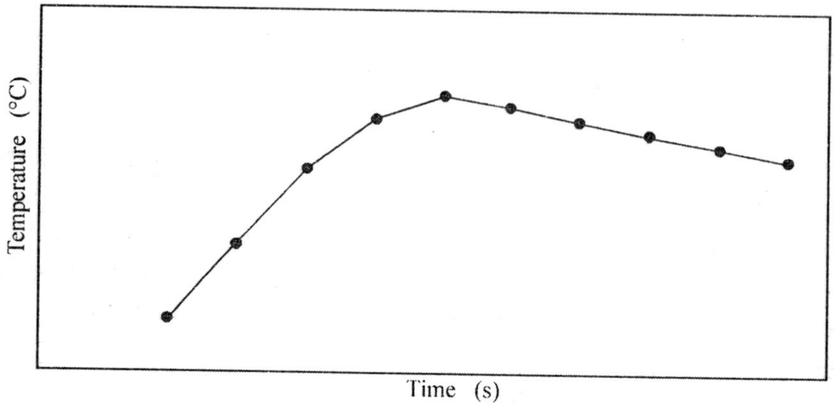

Fig. 3. Temperature-time graph

Temperature of the base taken $= t_4$°C =

Temperature attained after the addition of acid $= t_5$°C =

Heat generated, $\Delta H = (300 + M_1 s) \times (t_5 - t_4)$ = ... cals.

Result:

The heat of neutralization of HCl with NaOH = ...cal mol^{-1}

= .....J mol^{-1}

EXPERIMENT 12
Vapour Pressure and Enthalpy of Vapourisation of Water

Aim: To determine the vapour pressure of water at different temperatures and hence to evaluate the enthalpy of vapourisation of water.

Apparatus Required: A tall beaker, 10 cm^3 graduated cylinder, thermometer, Bunsen burner and graph sheet.

Principle: A sample of air is trapped in an inverted graduated cylinder immersed in a water bath. The trapped air is saturated with water vapour by heating the water to about 80°C. The volume of water vapour trapped inside the graduated cylinder is recorded. The bath is then allowed to cool and the volume of vapour recorded for every 5°C fall. The vapour pressure of water can be determined, knowing the barometric pressure. The number of moles of air in the mixture can be determined by measuring volume, and pressure at a temperature near 0°C, where the water content of the gas is negligible. The enthalpy of vapourisation is then calculated from Clausius- Clapeyron equation.

Procedure: Take about 6 cm^3 of distilled water in 10 cm^3 graduated cylinder; cover the top with a finger, invert the cylinder and quickly lower it into a tall beaker that has been filled with water, as shown in Fig. 4. Add more water to the tall beaker, if necessary, so that the inverted cylinder is fully immersed. Then, heat this set up on a Bunsen burner to approximately 80°C. Note the volume of the trapped air inside the inverted cylinder. On heating, the air expands. If it expands beyond the scale on the cylinder, remove the burner and allow the water to cool, so that the trapped air contracts. On cooling, note down the volume of the trapped air for every 5°C fall in temperature till the temperature reaches 50°C. (Small pieces of ice may be added to speed up the cooling process). After the temperature has reached 50°C, add ice to cool the water to a temperature lower than 50°C. Record this temperature and corresponding volume of the trapped air. Finally, record the atmospheric pressure at your laboratory from a barometer.

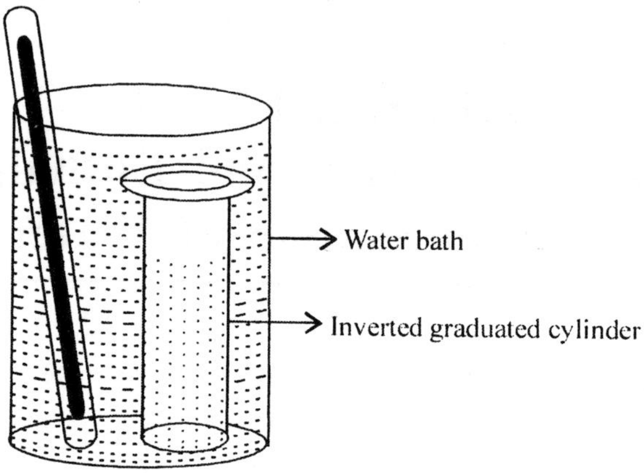

Fig. 4. Set up for vapour pressure measurement

Observations and Calculations:

1. Subtract 0.2 cm^3 from each volume measured from 80°C to 50°C, to compensate for the inverted meniscus. On inverting the graduated cylinder, the meniscus at the air-water interface is reversed. This reversal of meniscus would introduce a small systematic error. By introducing known volumes of air into a completely filled inverted cylinder, the error involved in inverting a 10 cm^3 cylinder has been found to be 0.2cm^3. This has to be subtracted to get the correct volume.
2. The number of moles of trapped air, n = PV/RT(assuming ideal behaviour), where P is the atmospheric pressure (noted from barometer), V is volume of the trapped air measured at a temperature (T) lower than 50°C and R being the gas constant.
3. Using the value of n calculated above, the partial pressure of air at each temperature can be obtained as

$$P_{air} = nRT/V_c$$

where V_c = corrected volume.
4. Then, the vapour pressure of water at each temperature can be found out, as it is given by

$$P_{H2O} = P_{atm} - P_{air}$$

Finally, tabulate the results obtained.

Tabulation: Number of moles of air trapped, n = ... correction factor = 0.2 cm^3

S. No.	Temp. T °C	Volume observed	Corrected volume (V_c)	Partial pressure of air, P_{air}	Vapour pressure of water, P_{H_2O}
1	80				
2	75				
3	70				
4	65				
5	60				
6	55				
7	50				

The Clausius – Clapeyron equation,

$$\log P_{H_2O} = -\Delta H_{vap}/2.303R\ (1/T) + C$$

Plot log P_{H_2O} values against 1/T and draw the best fit straight line. The slope of the line is = $-\Delta H_{vap}/2.303R$.

Hence, by measuring the slope of the straight line obtained in the above plot, the enthalpy of vapourisation of water, ΔH_{vap} can be calculated.

Result:

The enthalpy of vapourisation of water between 50°C and 80°C = . . . cal mol^{-1}. (The literature value of ΔH_v in this temperature range is 10.2 Kcal mol^{-1}).

PART IV

Phase Equilibria

EXPERIMENT 13
Transition Temperature of a Salt Hydrate

Aim: To determine the transition temperature of the given salt hydrate.

Apparatus Required: Pyrex test tubes, glass stirrer, thermometer, 400 cm^3 beaker.

Chemicals Required: Hydrated sodium sulphate or strontium sulphate or sodium thiosulphate.

Principle: There are many substances which are capable of existing in more than one crystalline form. In general, these different polymorphous forms are not equally stable at a given temperature. However, at a particular temperature, two different forms of the substance exist in equilibrium. This temperature is known as the transition temperature of that substance.

Not only are transition points associated with polymorphous substances but they are found in general, also with salt hydrates. When a salt combines with water to form one or more different hydrates, it is found that under a set of given conditions, only one of the hydrates of the salt is stable. Thus when sodium sulphate decahydrate is heated above 33°C, it decomposes into anhydrous sodium sulphate and a saturated solution of this salt. On the other hand, when a saturated solution of sodium sulphate is cooled below 33°C, the anhydrous salt takes up water and forms crystals of the decahydrate.

Similar relationships are observed with other salt hydrates also, e.g., sodium carbonate hydrates.

Procedure: Take a small amount of the given salt hydrate in the pyrex tube containing a glass stirrer and a thermometer. Add a small amount of liquid paraffin to the test tube to homogenize the system. Heat the contents of the tube slowly by means of a small flame using a water bath and record the rise in temperature for every minute, stirring the mixture continuously. Record the temperature of the mixture while cooling at regular intervals of time.

Treatment of Data:

Time (min)	1	2	3	4	5	6	7	8	9	10	11	12	13	14	15
Temp. on heating (°C)															
Temp. on cooling (°C)															

Plot the temperature versus time graph. Two curves (Fig. 5) with common horizontal portion will be obtained. The horizontal portion corresponds to the transition temperature (T_t) of the given salt hydrate.

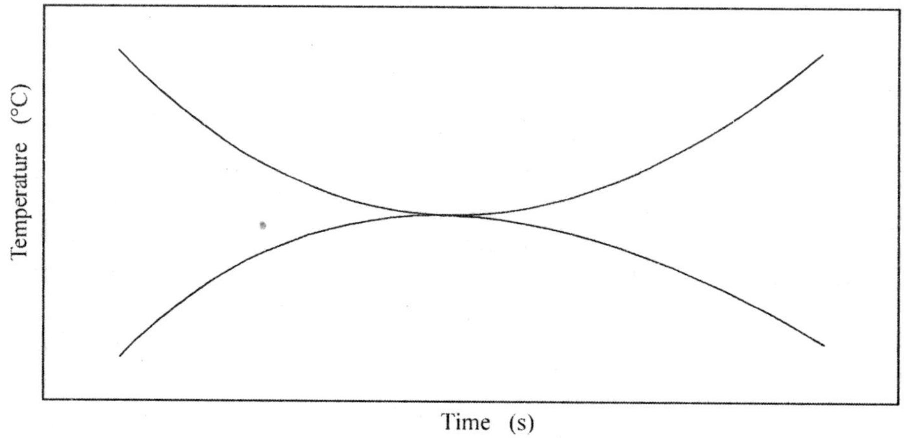

Fig. 5. Temperature-Time graph

Result:

The transition temperature of the given substance = . . . °C.

Table: Some typical transition temperatures of substances which can be tried for carrying out this experiment

Substance	Transition temperature °C
1. Ammonium nitrate	
Alpha rhombic to beta rhombic	32.3
Beta rhombic to rhombohedral	84
Rhombohedral to cubic	125
2. Potassium nitrate	129
3. Silver nitrate	159.6
4. Sulphur	95.5
5. Tin	13.2
6. Sodium sulphate	32.4
7. Mercuric iodide	126

EXPERIMENT 14
Critical Solution Temperature of Partially Miscible Liquid System

Aim: To determine the critical solution temperature of phenol – water system.

Apparatus Required: CST apparatus (pyrex test tubes provided with outer jackets and ring stirrer), thermometer, 400 cm^3 beaker, 100 cm^3 volumetric flask, 10 cm^3 burette.

Chemicals Required: Distilled phenol and distilled water.

Principle: Two liquids may be completly miscible or completely immiscible; between these extremes, there are systems consisting of two liquids which are partially miscible. For example, if a little phenol, ether or aniline is added to water at ordinary temperature, the material will dissolve completely, but if the addition is continued, a point is reached when no further dissolution occurs and two liquid layers are formed. One of these layers is a saturated solution of the organic substance in water and other is a saturated solution of water in the organic compound; the two systems in equilibrium are called 'conjugate solutions'. For a system of two components in which there are two liquid phases in equilibrium (the vapour being neglected since it is not in equilibrium), the phase rule gives,

$$F = C - P + 2 = 2 - 2 + 2 = 2$$

The system is bi-variant and so if the pressure is fixed at 1 atmosphere, the temperature alone will define the system completely. This means that at a given temperature, the compositions of the two liquid layers are fixed and are independent of the amounts of the two phases.

However, at a particular temperature the two liquids become completely miscible. This temperature, T_c is known as the 'critical solution temperature' or the consolute temperature of the system. Above this temperature, the two layers become miscible in all proportions.

Procedure: Pipette out 5 cm^3 of freshly distilled phenol into the inner tube of the CST apparatus (Fig. 6). Add water slowly from a burette till a persistent turbidity is seen. Note down the burette reading. Insert the thermometer into the inner tube and place it in a beaker of warm water and stir the liquid slowly by means of the stirrer and note the temperature at which the turbidity disappears. Take the apparatus out of the water bath and cool it slowly and again note the temperature at which turbidity just begins to appear. The mean temperature is taken as the miscibility temperature. In a similar manner, obtain the miscibility temperatures for various other mixtures made up of different proportions of phenol and water.

Treatment of Data: Using the density of phenol, convert the volumes into mass; calculate the mass percentage of phenol and tabulate the data in the following format.

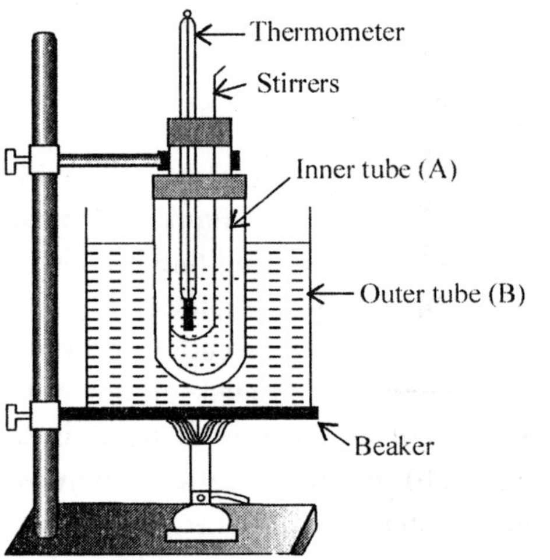

Fig. 6. The CST apparatus

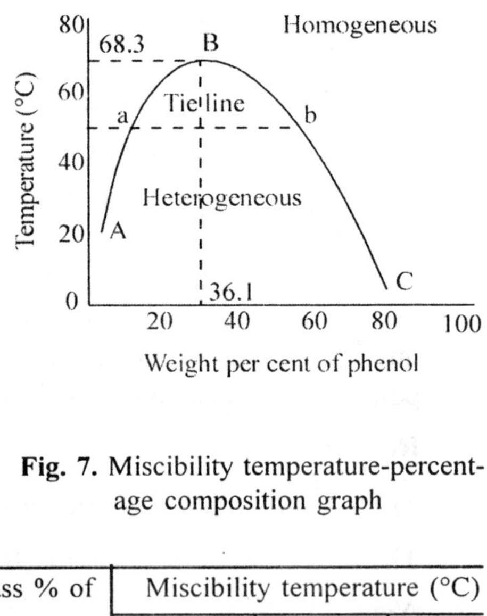

Fig. 7. Miscibility temperature-percentage composition graph

Volume of phenol in (cm^3)	Volume of water (cm^3)	Mass % of phenol	Miscibility temperature (°C)		
			Heating	Cooling	Mean
5	3.0	64.07			
5	4.0	57.22			
5	5.0	51.69			
5	5.5	49.31			

Volume of phenol in (cm^3)	Volume of water (cm^3)	Mass % of phenol	Miscibility temperature (°C)		
			Heating	Cooling	Mean
5	6.0	47.14			
5	6.5	45.15			
5	7.5	43.32			
5	7.5	41.62			
5	8.0	40.08			
5	8.5	38.63			
5	9.0	37.28			
5	9.5	36.03			
5	10.0	34.85			
5	11.0	32.72			
5	12.0	30.84			
5	13.0	39.16			
5	14.0	27.16			
5	15.0	26.28			
5	16.0	25.06			
5	17.0	23.94			

The % composition by mass of phenol is plotted on x-axis against the miscibility temperature on the y-axis (Fig. 6). The y-coordinate of the maximum point on the graph correspond to the critical solution temperature (T_c).

Result:

The critical solution temperature of phenol-water system = ...°C.

The composition of phenol at the critical solution temperature = ...%.

Note

1. It is not necessary for one of the substances to be water, as completely immiscible systems are frequently obtained with two organic compounds, e.g., aniline-hexane and carbon disulphide-methyl alcohol at ordinary temperature or with an organic compound and liquid sulphur or liquid phosphorus at higher temperatures.
2. For certain systems, as the temperature is lowered, the mutual solubilities increase and below a certain critical temperature the two

liquds become miscible in all proportions. This temperature is known as the 'lower critical solution temperature'. Certain systems can have both an upper and a lower consolute temperature e.g., nicotine-water system. Phenol–water system shows an upper CST behaviour.

3. Consolute temperature and composition for some common systems are given in the following table. (Lower critical solution temperature is exhibited by triethylamine–water system with its CST at 18.5°C).

System	Temperature (°C)	Composition (% of the first compound)
CH_3OH – cyclohexane	49.1	29
Isopentane-phenol	63.5	51
CS_2 – CH_3OH	40.5	80
Water – phenol	65.9	66
$AlBr_3$ – NaBr	232	92

EXPERIMENT 15
Effect of Electrolytes on the Critical Solution Temperature

Aim: To study the effect of added electrolyte on the critical solution temperature of phenol-water system and to determine the concentration of the given solution of the electrolyte.

Apparatus Required: CST apparatus (pyrex test tubes provided with outer jackets and ring stirrer), thermometer, 400 cm^3 beaker, 100 cm^3 volumetric flask, 10 cm^3 burette.

Chemicals Required: Distilled phenol and 2% NaCl solution.

Principle: The theoretical aspects of the simple phenol-water system has been dealt with already in the previous experiment. When a third substance is added to this system, at an impurity level, the mutual solubilities of the liquids depend on the nature and quantity of the added substance. If the substance is soluble in only one of the two liquids then it decreases the mutual solubilities and hence the consolute temperature is raised. On the other hand, when the added substance dissolves in both the liquids the consolute temperature is lowered (the addition of succinic acid to phenol-water system). Electrolytes, like NaCl are soluble only in water and not in organic liquids and hence they raise the CST due to the salting out effect. The increase in CST is proportional to the concentration of the added electrolyte. Therefore, by measuring the CST of phenol-water system at different concentrations of NaCl and forming the calibration graph, the concentration of the given solution can be read by interpolation.

Procedure: Mix 5 cm^3 of freshly distilled phenol and 9.0 cm^3 of distilled water in the inner jacket of the CST apparatus and find the miscibility temperature as described in the previous exercise. Then add 0.2 cm^3 of 2% NaCl solution to the above mixture of phenol and water and determine the miscibility temperature. Repeat the procedure by taking fresh phenol-water mixtures (but the same specified volume of phenol and water) for addition of 0.4, 0.6, 0.8 and 1.0 cm^3 of 2 % NaCl solution and as well as for the given NaCl test solution. Record the miscibility temperature in each case.

Treatment of Data: Record the miscibility temperature observed during heating and cooling in presence of different amounts of phenol and of NaCl in the following table. Use the data to plot a graph of mean miscibility temperature on the y-axis and concentration of NaCl added on the x-axis. A straight line graph will be obtained, (Fig. 8). Then the concentration of the given NaCl solution is obtained by interpolation.

Effect of Impurity: (Sodium chloride)

Volume of phenol (cm^3)	Volume of NaCl (cm^3)	Volume of water (cm^3)	% by weight of NaCl	Miscibility temperature, T°C		
				Heating	Cooling	Mean
5						
5						
5						
5						
5						
5						
5			(unknown)			

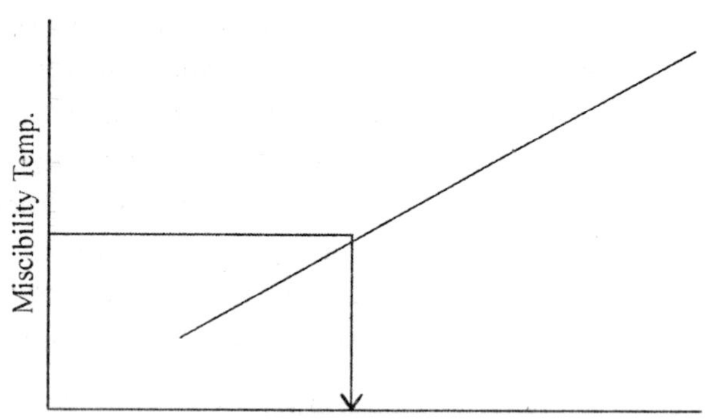

Fig. 8. Effect of [NaCl] on the miscibility temperature.

Result:

The concentration of the given NaCl solution = ...

Note: For accurate determination of the concentration, the effect of electrolyte (impurity) on the consolute temperature is to be studied by keeping composition of phenol (36.1%) corresponding to its CST. Also, to keep the composition of phenol constant through out the experiment, particular volumes of the electrolyte are to be added every time to 5 cm^3 of phenol and then diluted with water such that the total volume of the aqueous layer remain constant corresponding to that at the CST.

EXPERIMENT 16

Temperature-Composition Diagram for Two Component Liquid Azeotropic Systems

Aim: To obtain the temperature-composition phase diagram for a two component liquid system with maximum (or minimum) boiling point and to determine the maximum (or minimum) boiling point and composition.

Apparatus Required: Distilling unit, thermometer, dry test tubes, tripod stand, pipettes, Abbe's refractometer.

Principle: If the external pressure is maintained constant, the boiling point of a mixture of the two completely miscible liquids will vary with the composition. When the curve of total vapor pressure shows neither a maximum nor a minimum the boiling point curve will show neither a maximum nor a minimum and the boiling point of all mixtures will be intermediate between the boiling point of the pure components. The boiling point diagram for such systems, known as azeotropic mixtures is shown in (Fig. 9). When a solution which gives such a boiling point curve is distilled, the vapor (distillate) will be richer than the liquid in the component of higher vapor pressure and boiling point will rise as the distillation proceeds. By repeated fractional distillation of such a solution or by distillation through an efficient fractionating column, a complete separation of the two components may be effected.

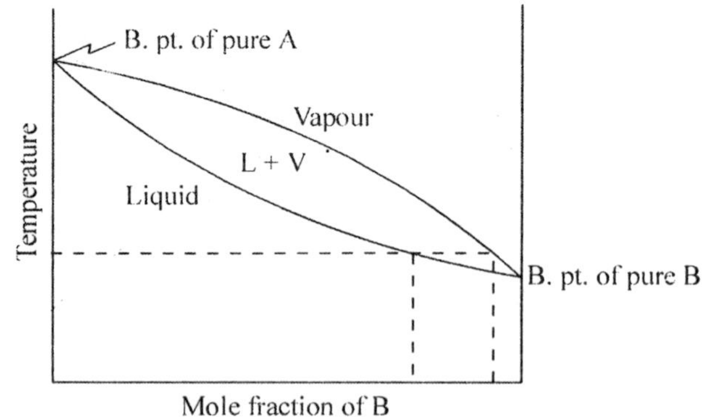

Fig. 9. Boiling point diagram of azeotropic mixture.

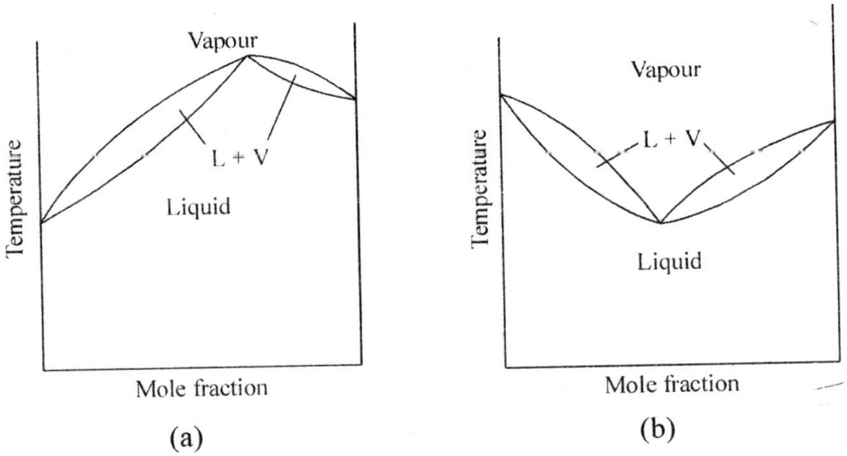

Fig. 10. Boiling point diagrams for azeotropic mixtures at (a) maximum and (b) minimum boiling points

When the vapor pressure curve of the liquid solution passes through a maximum, the boiling point curve will pass through a minimum and vice versa. In such cases, a separation of the mixture into its pure components cannot be effected by distillation, but only a separation into one or other of the components and a mixture of constant boiling point (azeotropic mixtures) the composition of which corresponds to the maximum or minimum point on the boiling point curve. The boiling point diagrams for azeotropic mixtures with maximum and minimum boiling points are given in (Fig.10).

Procedure: Prepare binary mixtures* containing 10, 20, 30, 40, 50, 60, 70, 80, and 90 mole % of one of the components.

1. Determine the refractive indices of each of these mixtures and that of the pure components using an Abbe refractometer.
2. Determination of the boiling point of the mixtures and the equilibrium composition of the vapor and the liquid:

Place the binary mixture in a distillation flask fitted with a thermometer and a condenser. The design of the special type of condenser to be used is shown in (Fig. 11).

Allow the liquid mixture in the flaks to boil steadily permitting the vapor to condense in the condenser. When the temperature of the liquid becomes constant, note the temperature. Stop heating and cool the distilling flask sufficiently. Remove the distillate and the residue in the flasks into a clean dry test tube by means of a dropper. Label them. Determine the refractive indices of the distillate and the residue using the Abbe's refractometer. Repeat the procedure for other binary mixtures.

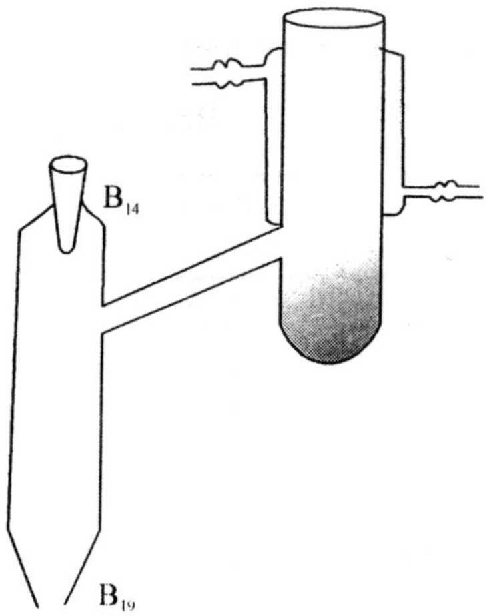

Fig. 11. Special type of condenser for collecting liquid samples for azeoptropic distillation.

Treatment of Data

1. *Refractive indices data :*

Molecular weight of component A = Density of component A =

Molecular weight of component B = Density of component B =

Refractive index of A = Refractive index of B =

Volume of component A in (cm^3)	Volume of component B (cm^3)	Mole per cent of A	Refractive index
		10	
		20	
		30	
		40	
		50	
		60	
		70	
		80	
		90	

Draw on a moderately large scale, a graph with composition of A on x-axis and refractive index on y-axis. A straight line graph will be obtained.

2. *Boiling point of the mixture and its composition:*

Boiling point of pure A = °C; Boiling point of pure B = °C.

Mole per cent of A	Boiling temperature (°C)	Refractive index		Mole per cent of A (from graph)	
		Distillate	Residue	Distillate	Residue
10					
20					
30					
40					
50					
60					
70					
80					
90					

Plot the composition of liquid (residue) and vapor (distillate) on x-axis against temperature on y-axis.

The given binary liquid system was found to be a two component system with minimum (or maximum) boiling point. The minimum (or maximum) point in the graph corresponds to the minimum/maximum boiling point and composition at this point is the composition of azeotrope.

Result:

The minimum (or maximum) boiling point of the two component liquid system = . . . °C.

The composition of the system at this temperature = . . .

---

*The components for the experiment may be suitably chosen from the systems given in the following table:

Azeotrope-two component mixtures

1. Maximum boiling point mixtures:

Component A	Component B	Constant boiling point (°C)	Mole % of A
Chloroform	Acetone	64.5	65.5
Nitric acid	Water	120.5	68.0**
Hydrochloric acid	Water	108.5	20.2**

**weight per cent of A in azeotrope

2. Minimum boiling point mixtures.

Component A	Component B	Constant boiling point (°C)	Mole % of A
Carbon tetrachloride	Methanol	55.7	44.5
Carbon tetrachloride	Ethanol	65.0	61.3
Carbon tetrachloride	n-propanol	72.8	75.0
Carbon tetrachloride	Isopropanol	67.3	64.0
Carbon tetrachloride	Methyl ethyl ketone	73.8	53.4
Carbon tetrachloride	t-butyl alcohol	69.5	71.0
Carbon tetrachloride	Ethyl acetate	74.8	43.0
Carbon tetrachloride	Isobutyl alcohol	75.8	89.0
Chloroform	Methanol	53.5	65.0
Chloroform	Ethanol	59.3	84.0
Chloroform	Isopropyl alcohol	60.8	92.0
Methanol	Acetone	55.7	20.0
Methanol	Methyl acetate	54.0	35.0
Methanol	Benzene	58.3	61.4
Methanol	n-hexane	50.6	51.0
Acetone	Methyl acetate	56.1	61.0
Isopropyl alcohol	Ethyl acetate	74.8	30.5
Isopropyl alcohol	Benzene	71.9	39.3
Ethanol	Water	78.15	95.6**
Acetic acid	Benzene	80.35	2.0**
Pyridine	Water	92.6	57.0**

**weight percent of A in azeotrope

EXPERIMENT 17

Two Component Simple Eutectic Systems

Aim: To trace the freezing point curve of two component simple eutectic systems.

Apparatus Required: Pyrex boiling tubes (3), glass stirrer, thermometer, 400 cm^3 beaker, tripod stand, wire gauze.

Principle: When a homogeneous liquid binary mixture is cooled, a solid eventually separates, but its composition depends entirely on the nature of the substances concerned. If the two components are chemically dissimilar and form no compound, then each lowers the freezing point of the other; they crystallize out separately and a eutectic (least melting point) system is obtained, having a minimum freezing point.

Procedure: Two substances A and B are given*. Weigh 5 g of A and 5 g of B. Then make ten packets each containing 0.5 g of B and 4 packets each containing 1 g of A. Determine the melting point of pure B taking 5 g of B. Then add 1 g of A to 5 g of B in a boiling tube fitted with a glass stirrer and a thermometer. Melt the mixture by immersing the tube in hot water, stirring the mixture to make the composition homogeneous. Cool the boiling tube (air cooling) containing the molten mixture and find out the temperature at which the first appearance of solid is perceptible. This will give the melting (freezing) point of the mixture. A is then added in steps of 1 g and the above procedure is repeated, after each addition.

In the same way, the melting point of pure A (5 g) is determined and B is added in steps of 0.5 g and the melting points of all the mixtures are determined as described above. For certain mixtures simple air cooling will not suffice, for the freezing point will be below the room temperature and in such cases the mixture has to be cooled in ice.

Obtain a temperature-composition graph (Fig. 12) by plotting the melting points of different mixtures against composition of one of the components taken on the x-axis.

The lowest point on the graph corresponds to the eutectic temperature and note the corresponding composition.

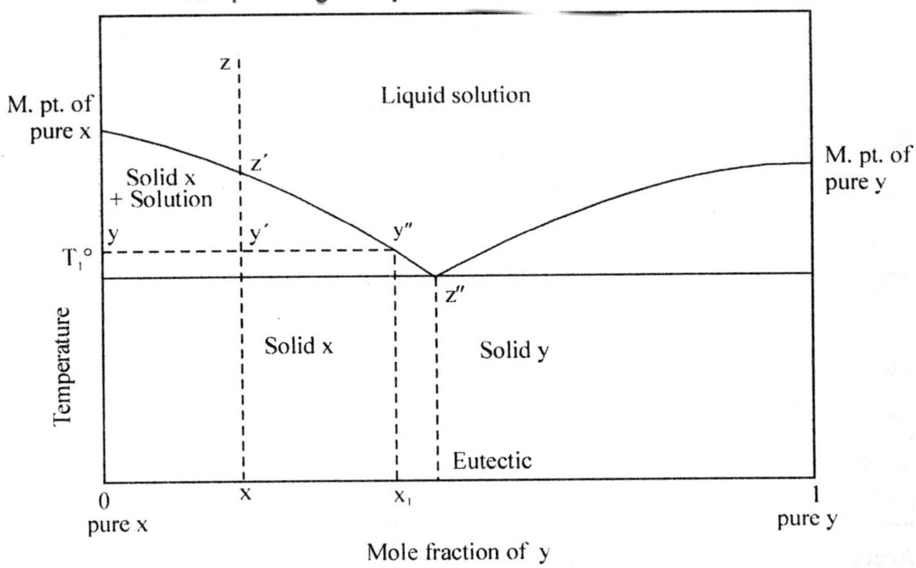

Fig. 12. Phase diagram for a typical binary mixture with simple Eutectic formation

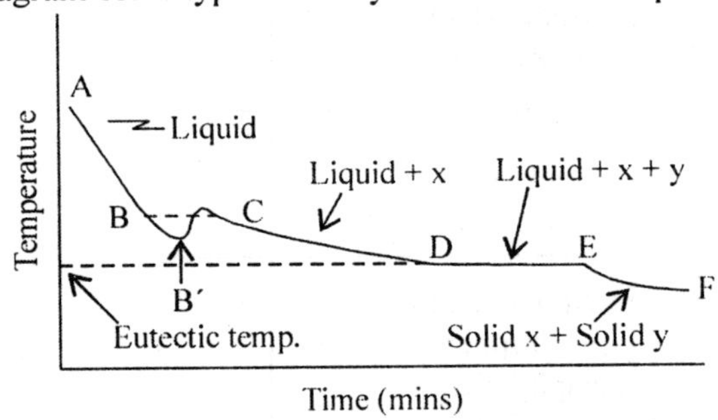

Fig. 12(a). Typical cooling curve for a binary mixture forming a eutectic.

Treatment of Data:

Weight of A (g)	Weight of B (g)	Percentage composition of A	Melting point (°C)
5.0	0.0		
5.0	0.5		
5.0	1.0		
5.0	1.5		
5.0	2.0		
0	5.0		

Weight of A (g)	Weight of B (g)	Percentage composition of A	Melting point (°C)
0.5	5.0		
1.0	5.0		
1.5	5.0		
2.0	5.0		

Results:

The eutectic temperature of the system...... = ...°C.

The eutectic composition of the system = . . . % of A and ... % of B.

**Note*: A and B may be suitably chosen from the systems given in the following table.

Two component solid-liquid systems

Component A melting point (°C)	Component B melting point (°C)	Eutectic temp. (°C)	Composition of B (mole %)
Acetamide (80.0)	Benzoic acid	38.0	45.0
Naphthalene (80.1)	Diphenyl (69.0)	39.4	56.0
Naphthalene	Diphenylamine (52.5)	30.5	63.6
o-nitrophenol (44.8)	Benzamide (124.8)	41.8	95 wt. % A

Component A melting point (°C)	Component B melting point (°C)	Eutectic temp. (°C)	Composition of B (mole %)
o-nitrophenol	Diphenylamine	21.0	51.0 wt. % A
Naphthalene	p-nitrophenol (113)	74.0	25.0
p-toluidine (43.5)	Naphthalene	31.9	27.5
p-nitro-toluene (53.0)	Naphthalene	29.0	36.5 wt. % B
p-nitrotoluene	Diphenylamine	18.5	49 wt. % A
picric acid (122.3)	m-dinitrobenzene (90)	61.0	62.4
o-nitrophenol	Naphthalene	29.0	34.0
Benzoic acid (121.5)	Cinnamic acid (136.0)	82.0	43.0
Benzoic acid	Naphthalene	70.0	67.0

EXPERIMENT 18
Two Component-Compound Forming Systems

Aim: To determine the freezing point curve of a two component - compound forming systems.

Apparatus Required: Pyrex boiling tubes with glass stirrer, thermometer, beakers, tripod stand, wire gauze.

Principle: When a homogeneous binary liquid mixture is cooled, a solid eventually separates, but its composition depends entirely on the nature of the substances concerned. If the two components are chemically dissimilar but form no compound, then each lowers the freezing point of the other having a minimum freezing point. If on the other hand, a stable compound (congruently melting) is formed, then the freezing point curve shows a maximum together with two eutectic minima. In some systems more than one compound is formed while others exhibit weak compound formation (incongruently melting compound).

If the components A and B form a solid compound AB, stable upto its melting point then the solid-liquid equilibrium diagram or the phase diagram will be of the type shown in Fig. 13. In addition to the two branches AC and BE which represent the compositions of liquid in equilibrium with solid A and solid B respectively, at different temperatures, there is a central portion CDE rising to a maximum. This portion of the curve represents liquid systems in equilibrium with the solid compound AB. It has been found experimentally, in accordance with theoretical requirements, that the maximum D of the curve occurs at the composition of the solid component as in this instance it is supposed to contain equimolar amounts of A and B the point D is midway between pure A and B.

At the point D, therefore liquid and solid phases have the same composition, so that D may be regarded as the melting point of the compound.

In this case the Compound AB is said to have congruent melting point, since solid and liquid of the same composition can coexist. At this point, the two component system has virtually become a one component system, so that the temperature at D is just as definite as the melting points of the pure components. The melting point of the compound may be above, below or between those of the two components. In this diagram, there are two eutectic points, viz., C and E at the former, the solids A and AB separate, whereas at the latter the solids B and AB separate.

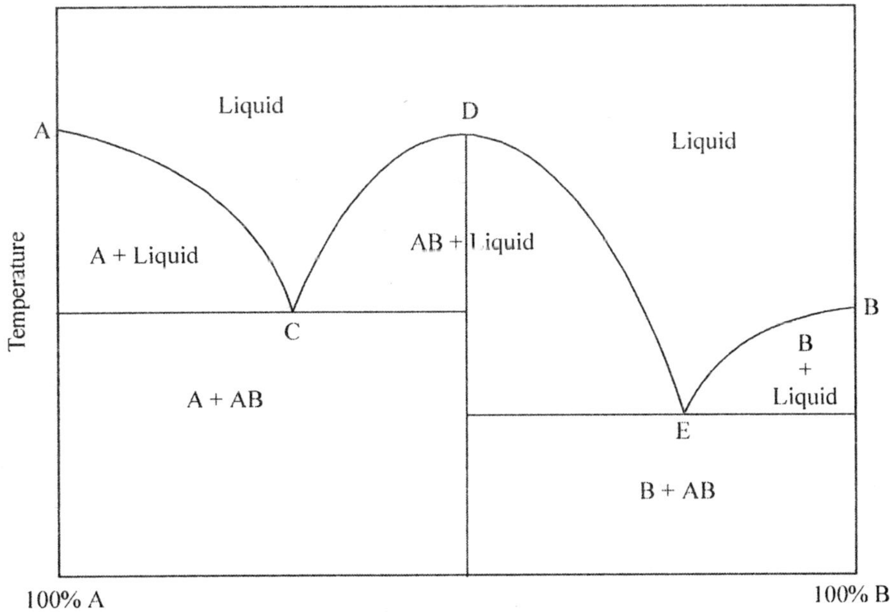

Fig. 13. Phase diagram for compound formation with congruent melting point.

Procedure: Weigh 5 g of A and determine its melting point. Make 8 packets of B each containing 0.5 g. Add the contents of one packet of B to 5 g of A taken in the tube fitted with a glass stirrer and thermometer. Melt the mixture by immersing the tube in hot water (liquid) bath and make the mixture homogeneous. Cool the boiling tube (air cooling may be sufficient) containing the molten mixture and find out the temperature at which the first appearance of solid is perceptible. This gives the melting point (or freezing point) of the mixture. Determine the melting points of all the other mixtures by adding the packets of B one by one,in the same manner.

Then weigh 5 g of B and determine its melting point. Then make four packets of A containing 0.5 g of A each and find out the melting points of the mixtures of A and B in an exactly analogous manner as above.

Observations and Calculation

Record the melting point of each mixture determined in the following table.

Weight of A (g)	Weight of B (g)	% composition by weight of B	Melting point (°C)
5.0	0.0		
5.0	0.5		
5.0	1.0		
5.0	1.5		
5.0	2.0		
0.0	5.0		
0.5	5.0		
1.0	5.0		
1.5	5.0		
2.0	5.0		

Draw a temperature-composition diagram by plotting the percentage composition by weight of B on the x-axis against the melting point of different mixtures on the y-axis.

The maximum point on the graph corresponds to the melting point of the compound AB. The other two minima on the graph correspond to the two eutectic temperatures.

Results:

1. The melting point of the compound AB is found to be.....°C and its composition is ....% by weight of B and ....% by weight of A.
2. The first eutectic point corresponds to a temperature of......°C and the eutectic composition is ....% by weight of B and ....% by weight of A.
3. The second eutectic point corresponds to a temperature . . . °C and the eutectic composition is ....% by weight of B and ....% by weight of A.

Two component solid-liquid systems (compound formation)

System melting point (°C)	Eutectic 1	Compound	Eutectic 2
1. A. Acetamide (80.6) B. β-naphthol (122.0)	53.2°C 52.0 mole % of A	63.0°C 30.0 mole % of A	61.4°C 25 mole % of A
2. A. Picric acid (122.3) B. Cannamic acid (13.0)	10.3°C 78.0 wt. % of A	106.5°C 60.7 wt. % of A	105.0°C 50.0 wt. % of A
3. A. Picric acid B. Naphthalene (80.1)	113.5°C 91.0 wt. % of A	149.0°C 64.1 wt. % of A	78.0°C 9.0 wt. % of A
4. A. Picric acid B. β-naphthol	111.0°C 91.0 mole % of A	157.0°C 50 mole % of A	116.0°C 6.0 mole % of A
5. A. salicylic acid (157.0) B. benzamide (124.8)	114.8°C 64 wt. % of A	116.0°C 53.3 wt. % of A	108.5°C 36.5 wt. % of A
6. A. p-toluidine (43.5) B. β-naphthol	38.2°C 89.0 wt. % of A	81.5°C 42.6 wt. % of A	78.0°C 33.5 wt. % of A
7. A. Diphenylamine (53.1) B. Benzophenone (47.8)	28.7°C 63.5 wt. % of A	30.85°C 48.1 wt. % of A	24.0°C 28.0 wt. % of A

EXPERIMENT 19
Three Component Liquid Systems: Acetic Acid – Benzene – Water

Aim: To study the phase diagram of a three component system.

Apparatus Required: 20 cm^3 burette (3), 250 cm^3 dry reagent bottles (6), 5 cm^3 pipettes.

Chemicals Required: Pure benzene (density = 0.88 g/cm^3), glacial acetic acid (density = 1.042 g/cm^3); a standard solution of sodium hydroxide and phenolphthalein indicator.

Principle: In three component systems a single phase possesses four degrees of freedom, viz., temperature, pressure and the composition of two of the three components. This number poses difficulty in graphical presentation of the phase relations. For this reason, data in ternary systems are generally presented at some fixed pressure, such as atmospheric pressure and at various constant temperatures. Under these conditions it is possible to show the concentration relations among the three components at any given temperature on a planar diagram. By combining such planar diagrams for various temperatures, it is then possible, if desired, to construct a solid model having concentrations as a base and temperature as a vertical axis.

For a three component system, phase rule takes the form $F = 5 - P$ (since $F = C - P + 2$). At a fixed pressure and temperature, the number of degrees of freedom is reduced by two, so that $F = 3 - P$ and the maximum number of phases which can occur simultaneously is thus three. This is the same number as is possible in two component systems under constant pressure only. Therefore, under the specified conditions, an area will again indicate bivariance, a line monovariance and a point, invariance.

Procedure

1. Titration of acetic acid – water mixture with benzene:

Prepare mixtures of acetic acid and water of different composition namely, 10%, 25%, 40%, 50%, 60%, and 80% of acetic acid in different bottles and label them properly. Place these stoppered bottles in a trough of water at laboratory temperature for about 15 minutes. Then remove the bottles and titrate the contents of each bottle against benzene taken in a burette. Shake the bottle vigourously after each addition of benzene. Note the volume of benzene required to bring about the first perceptible cloudiness.

2. Titration of acetic acid – benzene with water :

Prepare mixtures of acetic acid and benzene of different compositions, namely, 10%, 25%, 40%, 50%, 60% and 80% of acetic acid in different bottles and label them properly. Perform a similar experiment using these different mixtures, this time the contents of each bottle being titrated against water taken in the burette.

3. Data collection for the tie-line:

Prepare three different compositions of three components acetic acid, benzene and water. Pipette out 5 cm^3 of the benzene layer from one of the bottles into a weighed conical flask. Once again weigh the flask with the solution and note down the weight for the solution taken. Titrate the solution in the flask against standard alkali taken in the burette using phenolphthalein as the indicator. Similarly withdraw 5 cm^3 of the aqueous layer and carry out the titration. Repeat the above procedure for the other two compositions. From the titre values calculate the percentage composition of acetic acid in each of the layers.

Treatment of Data

1. Water-acetic acid mixture.

% acetic acid	Volume of benzene, (cm^3)
10	
25	
40	
50	
60	
80	

2. Benzene-acetic acid mixture.

% acetic acid	Volume of water, (cm^3)
10	
25	
40	
50	
60	
80	

3. Data of the tie line.

Weight % of acetic acid	Volume of acetic acid (cm^3)	Weight % of benzene	Volume of benzene (cm^3)	Weight % of water	Volume of water (cm^3)
10	4.8	45	25.68	45	22.5
20	9.6	45	25.68	35	17.5
30	14.4	45	25.68	25	12.5

4. Water-acetic acid mixtures versus benzene.

Volume of acetic acid (cm^3)	Volume of water (cm^3)	Volume of benzene (cm^3)	Per cent composition by weight of		
			Water	Acetic acid	Benzene
4.8	45				
12	37.5				
19.2	30				
24	25				
28.9	20				
38.4	10				

5. Benzene-acetic acid mixtures versus water.

Volume of acetic acid (cm^3)	Volume of benzene (cm^3)	Volume of water (cm^3)	Per cent composition by weight of		
			Acetic acid	Benzene	Water
4.8	41				
12	42.6				
19.2	34.1				
24	28.4				
28.9	22.7				
38.4	11.4				

1. Compute the percentage by weight of benzene, acetic acid and water for each of mixtures that showed the first indication of turbidity and plot them on a triangular graph paper. Thus the phase diagram for the three component system can be constructed. (For the method of plotting, see the instructions under the Note 2 given below).
2. Calculate percentage weight of acetic acid in each of the pairs of phases for the three mixtures. Plot these points on the triangular graph paper. Draw the lines passing through these points.

Results:

1. The three component phase diagram for the acetic acid-benzene-water system has been traced.
2. The tie lines for the compositions studied have been drawn.

***Note*:**

1. The ternary system of acetic acid-chloroform-water can also be employed to study the three component system characteristics.
2. Several schemes for plotting two dimensional equilibrium diagrams for ternary systems have been proposed. Of these the equilateral triangle method (see typical graph given in (Fig. 14)) suggested by Stokes and Roozeboom is most generally employed. In this method concentrations of the three components at any given temperature and pressure are plotted on an equilateral triangle. Each apex of the triangle is taken as 100 % of the component with which it is designated. If the vertices of the triangle are designated by A, B and C then to obtain percentages other than 100 for A the sides AB and

AC are divided into 100 (some time into 10) equal parts and lines parallel to the side BC are drawn. Each of these lines represents then a definite percentage of A ranging from zero on line BC to 100 at A. Similarly, lines dividing the sides BA and BC and parallel to AC give various percentages of B, while the lines along the sides CA and CB and parallel to AB represent various percentages of C.

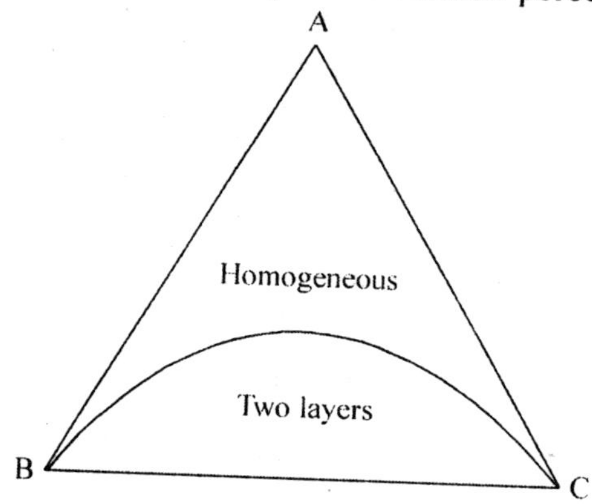

Fig. 14. Equilaletral triangle graph for three component systems.

3. Systems composed of three liquids which show partial miscibility may be classified as follows:

Type I : Formation of one pair of partially miscible liquids.

Type II : Formation of two pairs of partially miscible liquids.

Type III : Formation of three pairs of partially miscible liquids.

The three component systems of acetic acid-benzene-water (A-B-C) belong to type I with benzene and water forming one pair of partially miscible liquids. If we mix relative amounts of B and C so as to exceed the mutual solubility limits, two layers will be obtained; one composed of a solution of B in C, and the other of C in B. Suppose we add now to the two layers mixture a third liquid A, which is completely miscible with both B and C, A will distribute itself between the two layers and will promote thereby a greater miscibility of B and C. The increase in miscibility brought about by A depends on the amount of it added and on the amounts of B and C present. If sufficient A is introduced, the two layers can be changed into a single solution composed of the three liquids.

The Changes in miscibility produced by progressive additions of A to mixture of B and C can be followed on the diagram shown below. (Fig. 15)

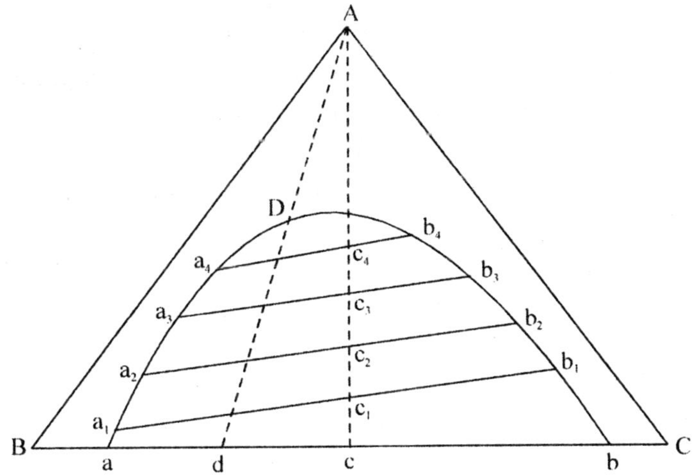

Fig. 15.

Points a and b designate the compositions of the two liquid layers resulting from mixing B and C in some arbitrary overall proportion such as c, while line Ac shows the manner in which the overall composition of c changed by addition of A. When enough A is added to change c to c_1, compositions of the two layers are shifted from a and b to a_1 and b_1 The line c_1 connects the compositions of the two layers in equilibrium and is called a tie-line. In a similar manner, the compositions reach c_2 and c_3. Finally at point b_4 sufficient A has been added to form only a single layer of this composition and there after only a single solution is obtained.

The tie-line for b_4 shows that the composition of B-rich layer at the point of complete miscibility is not identical with b_4 but is equal to a_4. This fact indicates that complete miscibility is brought about not by coalescence of the two layers into one, but rather by the diappearance of B-rich layer. Complete miscibility by merging of the two layers into one can occur only at one point on the diagram D. At this point the compositions of the two layers become identical and the two solutions coalesce into a single liquid phase of constant composition. Point D is called either the isothermal critical point of the system or the plait point and can be obtained only by adding A to a single mixture of B and C namely d.

Any mixture of A, B and C of overall composition falling within the same shaped area will yield two liquid layers of compositions given by appropriate tie-line through the composition of the mixture. On the other hand, any mixture of overall composition outside this area will yield only a single homogeneous solution of the three liquids. Curve aDb is frequently referred to as bimodal curve. In general, the plait point D will lie off the maximum of bimodal curve. Furthermore, since as a rule component A will not distribute itself equally between the two layers a and b, the tie-lines will not be parallel either to BC or to each other.

PART V

Chemical Kinetics

EXPERIMENT 20
Acid-Catalysed Hydrolysis of Methylacetate

Aim: To study the kinetics of hydrolysis of methylacetate in presence of a mineral acid and to evaluate the value of the rate constants.

Apparatus Required: Conical flasks, standard flasks, burette, pipettes and stop watch.

Chemicals Required: Distilled methylacetatc (or ethyl acetate), 0.5 N HCl, 0.2 N NaOH, phenolphthalein indicator, ice cold distilled water.

Principle: The overall reaction can be represented as

$$CH_3-\overset{\overset{\displaystyle O}{\|}}{C}-OCH_3 + H_2O + H^+ \longrightarrow CH_3COOH + CH_3OH$$

The rate of this reaction is given by, rate = k[ester] $[H_2O]$ $[H^+]$.

However, the effective $[H_2O]$ and $[H^+]$ would remain constant throughout the reaction, since they are used as solvent and catalyst respectively.

The rate equation under the conditions specified above can be written as

$$\text{Rate} = k'[\text{ester}]$$

where

$$k' = k[H_2O]\,[H^+]$$

Here, k and k′ are respectively the true and apparent rate constants for the reaction. Under these conditions it is said to be a pseudo first order reaction and the value of the its rate constant is given by the expression

$$k' = (2.303/t) \times \log\{a/(a - x)\}$$

where a and (a – x) are respectively the initial concentration and the concentration of the ester remaining at time t.

Procedure: Pipette out 100 cm^3 of the 0.5 M hydrochloric acid in a 250 cm^3 stoppered bottle. Keep the bottle in a trough of water at laboratory temperature (note the temperature) so that the contents of the bottle attain the temperature of the bath. Pipette out 2 cm^3 of the ester into the stoppered bottle. Shake the bottle to ensure thorough mixing. Immediately withdraw, by means of a pipette, 5 cm^3 of the reaction mixture and transfer it into a conical flask containing 20 cm^3 of ice cold water. Add a drop of phenolphthalein to the conical flask and titrate against standard (0.2 N) sodium hydroxide solution taken in the burette. The first appearance of pale pink colour is the end point. Note down the titre value (V_0). This value corresponds to time, t = 0.

At definite intervals of time, say 10, 20, 30, 40, 60, 80 minutes pipette out 5 cm^3 of the reaction mixture, deliver the same into ice-cold distilled water (20 cm^3) in a conical flask and titrate against the standard sodium hydroxide solution. Note the titre value (V_t) in each case. Keep the bottle (loosely stoppered) in a water bath maintained at 80°C for 15 minutes. This enables the entire ester to hydrolyse yielding an equivalent quantity of acetic acid. Take precaution that the stopper would not pop off while heating to 80°C. Cool the contents of the bottle, withdraw 5 cm^3 of the reaction mixture into the conical flask containing 20 cm^3 of ice-cold distilled water and add phenolphthalein indicator and titrate against the standard sodium hydroxide. Note the titre value (V_∞).

Treatment of Data: Acid hydrolysis of the ester at temperature = . . . °C = . . . k.

($V_\infty - V_0$) corresponds to the initial concentration (a) of the ester taken, ($V_\infty - V_t$) corresponds to the concentration of the ester at any time i.e., (a – x). The kinetic equation in terms of titre values reduces to

$$k = (2.303/\ t) \times \log [(V_\infty - V_0)/(V_\infty - Vt)]$$

Using this equation calculate the rate constant.

The titre values are used instead of concentration since they are directly proportional to concentrations. A plot of log ($V_\infty - V_t$) against t would yield a straight line with its slope equal to – 2.303/k from which k can be evaluated.

Time (min.)	Burette reading (cm^3)	Titre (cm^3)	Value	Rate constant k (s^{-1})
0				
5				
10				
20				
30				
40				
60				
∞				

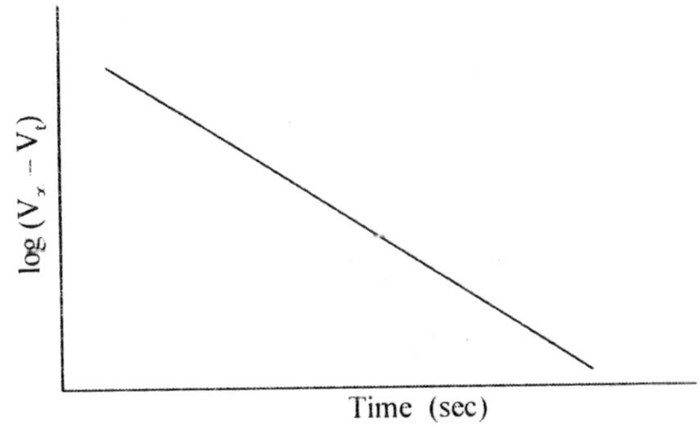

Fig. 16. First order plot

Results:

The rate constant for the acid catalysed hydrolysis of the given ester at ......°C. is

(i) by calculation, k = ...s^{-1}

(ii) from the graph, k = ...s^{-1}

Note:

1. The unit of the k is $time^{-1}$ usually s^{-1} or min^{-1}.
2. It may be noted that the strength of the sodium hydroxide used for the titration need not be known, as long as the reaction mixture is titrated against the same solution of sodium hydroxide. Plot log $(V_\infty - V_t)$ values against time and measure the slope. From the slope calculate the value for the rate constant.

EXPERIMENT 21
Alkaline Hydrolysis of Ethylacetate by Volumetry

Aim: To determine the specific rate of the alkaline hydrolysis of ethylacetate by volumetric method.

Apparatus Required: 50 cm^3 burette; 50, 10 and 5 cm^3 pipettes, conical flasks, stoppered bottles, porcelain trough, thermostat.

Chemicals Required: 250 cm^3 of ethylacetate solution (M/60), 500 cm^3 of sodium hydroxide solution (M/40), Sodium hydroxide solution (0.02 M), Hydrochloric acid (0.05 M), Phenolphthalein indicator.

Principle: The reaction is represented as

$$CH_3\ COOC_2H_5 + NaOH \longrightarrow CH_3\ COONa + C_2H_5OH$$

This reaction follows second order kinetics. After mixing the ester and NaOH, aliquots are withdrawn at different time intervals and unreacted NaOH is neutralised by adding a known excess quantity of acid. The unreacted acid is then found out by titrating it against a standard base. From these titre values, the second order rate constant can be evaluated using the formula,

$$k_2 = \{2.303/t(a - b)\}\ \log\{b(a - x)/a(b - x)\}$$

where a and b are the initial concentrations of ester and alkali respectively, x is the amount of ester reacted and t is the time.

Procedure: 50 cm^3 of the ester solution is pipetted out into a 250 cm^3 stoppered bottle. In another 250 cm^3 stoppered bottle pipette out 50 cm^3 of M/40 NaOH solution and thermostat both the bottles for nearly 15 minutes to attain constant temperature; note the temperature. After they have attained constant temperature, mix both the solutions and start a stopwatch. As soon as the mixing is carried out, pipette out 10 cm^3 of the reaction mixture into a conical flask containing 10 cm^3 of 0.05 M HCl. This will neutralise the unreacted alkali and thus arrest the hydrolysis of the ester. (Note: why acid catalysed hydrolysis cannot affect the reaction? – compare at the end of the experiment the values of the rate constants under both, acid - catalysed and alkaline hydrolysis and justify for yourself). The contents of the conical flask is now titrated against standard (0.02 M) alkali taken in the burette. The procedure of withdrawing 10 cm^3 portions of the reaction mixture, arresting the reaction with the use of standard acid of known concentration and known amount and titrating the contents of the conical flask is repeated at various time intervals of 3,5,10,15,20,30,40,50 minutes. Now keep the contents of the remaining reaction mixture in a temperature bath maintained at 80°C for about 15 minutes. Cool the solution, pipette out 10 cm^3 of the reaction mixture into a conical flask containing 10 cm^3 of 0.05 M HCl. Titrate the contents of the conical flask against the same standard alkali.

Treatment of Data: Tabulate the data obtained as given below:

Volume of alkali required to neutralize the amount of HCl taken for quenching (T_0) = ...cm^3,

Volume of alkali required to neutralize HCl at infinite time (T_∞) = ...cm^3.

Concentration of the alkali used = ... N;

Volume of the aliquot (V) = ...cm^3.

Time (min)	Titre (T_t) (cm^3)	$T_0 - T_t$	$T_\infty - T_t$	$\log [(T_\infty - T_i)(T_0 - T_t)/(T_0 - T_i)(T_\infty - T_t)]$	k_2
3					
5					
10					
15					
20					
30					
40					
50					

Mean, k_2 = ...

Let T_0 is the amount of the alkali required to neutralize the total acid used every time to quench the reaction, T_I is the first titre value (base required to neutralize the acid used for quenching at the start of the reaction) and T_t is the titre value at time t and T_∞, the titre value at infinite time, one can write the value of the second order rate constant in terms of titre values as follows:

$$k_2 = [\{2.303/t(T_0 - T_\infty)\} \log \{(T_\infty - T_1)(T_0 - T_t)/(T_0 - T_i)(T_\infty - T_t)\}]/(N/V)$$

where N is the concentration of the alkali used (in normality) for titration and V is the volume of the aliquot taken every time. These last two factors convert the titre values to concentration units especially in gram equivalents per litre which is equal to mole per litre in this case; thus the value of the rate constant is in the units of concentration^{-1} time^{-1}.

Result:

The value of the second order rate constant for the alkaline hydrolysis of ethylacetate at ....^{o}C is k_2 = ...lit.mol^{-1} s^{-1}.

EXPERIMENT 22
Alkaline Hydrolysis of Ethylacetate by Conductometry

Aim: To determine the second order rate constant for the alkaline hydrolysis of ethylacetate by conductivity method.

Apparatus Required: Conductivity cell, conductivity bridge, beakers 50 cm^3, pipettes 25 cm^3, stop watch, thermostat.

Chemicals Required: N/20 NaOH, N/5 ethylacetate, N/100 sodiumacetate, (all prepared in conductivity water).

Principle: The overall reaction can be given as

$$CH_3COOC_2H_5 + NaOH \longrightarrow CH_3COONa + C_2H_5OH.$$

The reaction is followed by the change in conductance of the reaction mixture at different time intervals. As no indicator is required for these titrations, this method is simple and easy to follow.

Procedure: Measure the resistance of an N/100 NaOH solution taken in a beaker. Record this value as R_0. Measure the resistance of an N/100 sodiumacetate solution taken in a beaker and record it as R_a. Pipette out 20 cm^3 of N/20 NaOH into a 250 cm^3 beaker and add 50 cm^3 of conductivity water and keep it in a thermostat. Pipette out 10 cm^3 of N/5 ethylacetate into another beaker and add 50 cm^3 of conductivity water and keep this beaker also in a thermostat. After thermal equilibrium is attained, pipette out 30 cm^3 of the solution containing the ester into the beaker containing the sodium hydroxide and start a stop watch. Mix the solutions and place the conductivity cell in the beaker and record the resistance (R_t) of the solution at regular time intervals until the reaction is completed.

Treatment of Data:

Table: Values of resistance of the reaction mixture at various time intervals.

Time (min)	Resistance (R_t) (ohm)	Conductance (C_t) (mho)	$k_2 = \{(C_0 - C_t)/(C_t - C_a)\}/\{0.01 \times t\}$

C_o and C_a are the conductance of N/100 NaOH solution (computed from R_0) and of N/100 sodiumacetate solution (calculated from R_a value).

Draw a graph by taking $(C_o - C_t)/(C_t - C_a)$ on the X-axis and t on the Y-axis. Find out the value of the rate constant from the slope of the best fit line drawn.

Result:

The value of the rate constant k_2 = ... L $mol^{-1}s^{-1}$ by calculation and

k_2 = ... L $mol^{-1}s^{-1}$ from the graph.

EXPERIMENT 23
Activation Energy for the Acid - Catalysed Hydrolysis of Methylacetate

Aim: To study the kinetics of the acid-catalysed hydrolysis of methylacetate and hence to evaluate the activation energy for the reaction.

Apparatus Required: Conical flask, volumetric flasks, 50 cm^3 burette, 5 cm^3 and 2 cm^3 pipettes, stop watch and thermostat.

Chemicals Required: Distilled methylacetate, 0.5 N HCl, 0.2 N NaOH, phenolphthalein indicator, distilled water and ice-cold distilled water.

Principle: The overall reaction is represented as,

$$CH_3COOCH_3 + H_2O + H^+ \longrightarrow CH_3COOH + CH_3OH$$

It is a pseudo first order reaction and follows the rate law, rate = k [ester], where k is equal to $k'[H_2O]\,[H^+]$, k and k′ being the true and the apparent rate constants respectively. The rate constant, k′ is given by $k' = (2.303/t) \log \{a/(a - x)\}$ where a and (a – x) are the concentrations of the ester at zero and at a given time 't' from the start of the reaction.

The activation energy is the energy required to form the activated complex from the reactants. This activation energy, E_a, is related to rate constant by the Arrhenius equation as, $\log k = \log A - E_a / (2.303 \times R \times T)$.

Therefore, from the slope of the straight line generated by plotting the log k values against 1/T, the value of E_a can be obtained.

Procedure: Pipette out 100 cm^3 of the 0.5 N HCl solution in each of three different 250 cm^3 stoppered reaction bottles. Thermostat these bottles at 30,35 and 40°C at three different thermostats, each of which has been already set for that particular temperature. (one of the bottles may be kept in a trough of water, instead of thermostating at 30°C, and the trough temperature may be noted and used for calculations later). In a separate bottle, thermostat the methylacetate sample also at the respective temperatures. After the solutions attained thermal equilibrium (say, in about 20 min.), pipette out 2.0 cm^3 of the ester from the respective thermostat, into the corresponding reaction bottle containing 0.5 N HCl. When the pipette is half-drained, start the stop watch. Shake the bottles thoroughly for uniform mixing. Immediately, withdraw, 5 cm^3 of the reaction mixture using a pipette and transfer it into a conical flask containing about 20 cm^3 of ice cold distilled water. Add a drop of phenolphthalein indicator and titrate it against the 0.2 N NaOH solution taken in the burette. The first appearance of pale permanent pink colour is the end point. Note the titre value as V_0. This corresponds to time, t = 0. At definite time intervals, say 10,20,30,40, etc, minutes, pipette out 5 cm^3 aliquots of the reaction mixture from bottles kept at different temperatures, into different conical flask, quench the reaction in ice cold water and titrate against the same 0.2 N NaOH as before. Note the titre values (V_t) in each case. Then, keep the bottles (loosely stoppered) in a water bath maintained at 80°C for 15 minutes. Cool the contents, pipette out 5 cm^3 of this mixture into a conical flask; add phenolphthalein indicator and titrate against 0.2 N NaOH. This titre value corresponds to V_∞.

Treatment of Data:

I. Acid catalysed hydrolysis of methylacetate at 30°C

V_0 = ...cm^3; V_∞ = ... cm^3; $V_\infty - V_0$ = ... cm^3

Time (s)	Burette reading		Titre value (V_t) (cm^3)	($V_\infty - V_t$) (cm^3)	log ($V_\infty - V_t$)
	Initial	Final			
600					
1200					
1800					
2400					
3000					
3600					

II. Acid catalysed hydrolysis of methyl acetate at 35°C

V_0 = ...cm^3; V_∞ = ... cm^3; $V_\infty - V_0$ = ... cm^3

Time (s)	Burette reading		Titre value (V_t) (cm^3)	($V_\infty - V_t$) (cm^3)	log ($V_\infty - V_t$)
	Initial	Final			
600					
1200					
1800					
2400					
3000					
3600					

II. Acid catalysed hydrolysis of methylacetate at 40°C

V_0 = ……cm^3; V_∞ = … cm^3; $V_\infty - V_0$ = ……… cm^3

Time (s)	Burette reading		Titre value (V_t) (cm^3)	($V_\infty - V_t$) (cm^3)	log ($V_\infty - V_t$)
	Initial	Final			
600					
1200					
1800					
2400					
3000					
3600					

For the data at each temperature, plot log ($V_\infty - V_t$) values against time. From the slope of the line obtained calculate the value for the rate constant, k, and tabulate the values as shown below:

Temperature		1/T	k s^{-1}	log k
°C	K			
30	303			
35	308			
40	313			

Plot log k values against 1/T (in K). Measure the slope of the straight line obtained. Calculate E_a, as

$$E_a = 2.303 \times R \times \text{slope}$$

(where R = 8.314 $JK^{-1} mol^{-1}$)

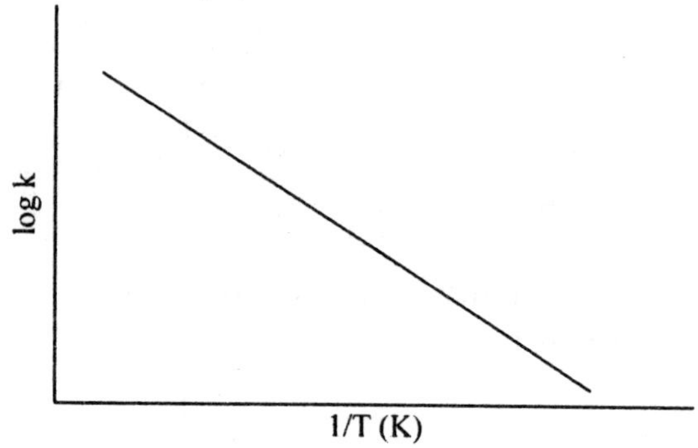

Fig. 17. Arrhenius plot

Extrapolate the straight line and find out the y- intercept, which gives the log A value, from which the frequency factor, A can be computed.

Results:

(i) The rate constant for the acid catalysed hydrolysis of methyl acetate
At 30°C = ...s^{-1}; At 35°C = ...s^{-1}
At 40°C = ...s^{-1}

(ii) The energy of activation for the reaction is
E_a = ...J

(iii) The value of the frequency factor = ...

EXPERIMENT 24
Dissociation of Trichloroacetic Acid – Kinetic Method

Aim: To determine the dissociation constant of trichloroacetic acid from the kinetic data of hydrolysis of an ester.

Apparatus Required: 50 cm^3 burette, 250 cm^3 stoppered bottles, 250 cm^3 conical flasks, pipettes.

Chemicals Required: 0.2 M sodium hydroxide, pure methyl or ethyl acetate 0.25 M and 0.5 M hydrochloric acid, 1.0 M trichloroacetic acid, Phenolphthalein indicator.

Principle: The kinetics of acid - catalysed hydrolysis of an ester is followed at fixed concentration of a strong acid (HCl) catalyst and also in the presence of trichloroacetic acid as the catalyst. The latter being relatively weak, ionizes only partially.

$$CCl_3COOH \rightleftharpoons CCl_3COO^- + H^+$$

Therefore, the ratio of $[H^+]$ to $[CCl_3COOH]$ taken initially would correspond to the degree of dissociation, α.

The overall reaction of ester hydrolysis is,

$$R - COOR' + H_2O + H^+ \longrightarrow RCOOH + R'OH$$

The observed rate constant is given by

$$k_{apparent} = k_{true}\,[H^+]$$

Using the $k_{apparant}$ and the k_{true} values in presence of HCl and CCl_3COH, the $[H^+]$ and hence α value is calculated, from which the dissociation constant is then arrived at as given in the latter part of this experiment.

Procedure: This experiment is carried out similar to the procedure adopted for the hydrolysis of an ester in presence of an acid. Take 50 cm^3 of 0.5 M HCl in a 250 cm^3 stoppered bottle and keep it in a trough of water so that it attains a constant temperature. Add 2 cm^3 of the ester, starting a stopwatch simultaneously. Immediately withdraw 5 cm^3 of the reaction mixture into a conical flask containing ice cold water and titrate the contents against standard alkali solution using phenolphthalein as indicator. Repeat the withdrawal of 5 cm^3 aliquots at regular time intervals of say 5, 10, 15......min. and titrate with the alkali. These corresponds to V_t. As well as find the titre value at infinite time for a 5 cm^3 aliquot of the reaction mixture, after heating the reaction mixture at 80°C for 15 minutes followed by cooling. This is V_∞ value.

Repeat this experiment with another concentration of HCl, say 0.25 M. Proceed exactly the same way with 0.5 M or 1.0 M of the given trichloroacetic acid to collect the kinetic data for the hydrolysis of the same ester. Carry out blank (without ester) titrations for each of these acid concentrations. This titre value is taken as V_0 value in each case.

Treatment of Data:

I. Kinetic data for 0.5 M HCl-catalysed reaction:

$V_0 = \ldots\ cm^3$; $V_\infty = \ldots\ cm^3$

S. No.	Time (s)	Titre (V_t) (cm^3)	$\log [(V_\infty - V_0)/(V_\infty - V_t)]$	k s^{-1}
1				
2				
3				
4				
5				
6				

Mean k = ... s^{-1}. (This k is $k_{apparent}$ at 0.5 M HCl.)

∴ $k^1_{true} = k_{apparanent}/0.5 = \ldots$

II. Kinetic data for 0.25 M HCl-catalysed reaction:

$V_0 = \ldots\ cm^3$; $V_\infty = \ldots\ cm^3$

S. No.	Time (s)	Titre (V_t) (cm^3)	$\log [(V_\infty - V_0)/(V_\infty - V_t)]$	k s^{-1}
1				
2				
3				
4				
5				
6				

Mean k = ... s^{-1}.

∴ $k^1_{true} = k_{apparanent}/0.5 = \ldots$

∴ Mean: $(k^1_{true} + k^2_{true})/2 = k_{true}$.

III. Data for 0.5 M CCl_3COOH-catalysed reaction:

V_0 = ...cm^3; V_∞ = ...cm^3

S. No.	Time (s)	Titre (V_t) (cm^3)	$\log [(V_\infty - V_0)/(V_\infty - V_t)]$	$k\ s^{-1}$
1				
2				
3				
4				
5				
6				

Mean k = ... s^{-1}.

Calculate the value of the specific rate constants numerically using the equation,

$$k = (2.303/t) \log\{(V_\infty - V_0)/(V_\infty - V_t)\}$$

as well as graphically for 0.5 M, 0.25 M HCl as well as for 0.5 M trichloroacetic acid. Calculate the value of the true rate constant of the reaction using the relation $k_{apparent} - k_{true} [H^+]$ from the data obtained with HCl.

The hydrogen ion concentration is therefore given by = $[H^+] = k_{apparent}/k_{true}$.

Here $k_{apparent}$ is the rate constant obtained when trichloroacetic acid is used as the catalyst and k_{true} is the mean of the rate constants calculated from those obtained with HCl as the catalyst. Calculate the degree of dissociation as follows:

$$\alpha = [H^+]/[\text{concentration of acid used}]$$

Calculate the value of the dissociation constant using the relation,

$$K = a^2c/1 - \alpha$$

Where c is the concentration of the acid used.

∴ $k_{apparent} / k_{true} = [H^+]$, liberated from CCl_3COOH.

∴ $\alpha = [H^+]/[CCl_3COOH]_{initial} = [H^+]/0.5$

The dissociation constant, is then calculated as, $K_a = \alpha^2c/1 - \alpha$

Result:

The dissociation constant of tricholoroacetic acid = ...

EXPERIMENT 25
Hydrolysis of t-amyl iodide

Aim: To evaluate the rate constant for the hydrolysis of t-amyl iodide at different initial concentrations of the reactant as well as to evaluate the value of the rate constant when 80 and 60% aqueous ethyl alcohol is used as solvent.

Apparatus Required: Thermostat, conductivity bridge, conductivity cell, 250 cm^3 stoppered conical flask, 400,250 ad 50 cm^3 beakers, graduated 1 cm^3 pipette, 5 cm^3 pipette and 100 cm^3 measuring cylinder.

Chemicals Required: Ethyl alcohol, t-amyl iodide.

Principle: The hydrolysis of t-amyl iodide in a solvent such as aqueous ethanol is a first order unimolecular reaction; the rate determining step of this reaction is the dissociation step in the reaction sequence

$$t - C_5H_{11}I \xrightarrow{\text{slow}} t - C_5H_{11}^+ + I$$

$$t - C_5H_{11}^+ + H_2O \xrightarrow{\text{fast}} t - C_5H_{11}OH + H^+$$

Since ions are formed as a result of hydrolysis, the reaction kinetics can be followed conductometrically. For a first order reaction, the half life of the reaction is independent of the initial concentration of the reactant used. In addition, the rate constant of reactions wherein ions are formed from covalent molecules is increased with increase in the polar nature of the solvent used.

Procedure: Prepare 500 cm^3 of aqueous ethyl alcohol solution containing 80 % by volume of ethyl alcohol. Pipette out 250 cm^3 of this solution into a stoppered bottle and thermostat the same. When the thermal equilibrium is reached, pipette out 50 cm^3 of the aqueous alcohol into a beaker and immerse the conductivity cell into the same. Pipette out 0.3 cm^3 of the t-amyl iodide into the beaker and start a stopwatch. Stir the contents of the beaker vigorously. Measure the conductance of the solution as soon as possible. Take conductance readings at half minute intervals for the first 5 minutes and at one minute intervals for a further ten minutes. To find the conductance (infinite value) of the solution after the completion of the hydrolysis, place the reaction mixture in a thermostat at 60-65°C for about 10 minutes and then cool the solution and record the conductance value. Repeat the experiment with 0.15 and 0.45 cm^3 of tertiary amyl iodide.

Repeat the experiment using 0.3 cm^3 of t-amyl iodide but using a different aqueous ethyl alcohol (say 60 % or 70 %).

Treatment of Data: Tabulate the observed conductance data as follows:
t-amyl iodide = 0.3 cm^3

Time (s)	30	60	90	120	150	180	210	240	270	300	330	360	390	∞
80% ethanol														
70% ethanol														
60% ethanol														

Repeat this tabulation for 0.15 and 0.45 cm^3 of t-amyl iodide. Plot on a large graph paper observed conductance values as a function of time. From the graph read off the time ($t_{1/2}$) required for 50 % completion of the reaction.

Calculate the value of the rate constant as $k = 0.693/t_{1/2}$ where $t_{1/2}$ is the half life of the reaction and k, the rate constant.

Results:

(i) Report the value of the rate constant under specific conditions of the reaction.

(ii) Comment on the $t_{1/2}$ value when different initial concentrations of t-amyl iodide was used.

(iii) Deduce the order with respect to the [t-amyl iodide]

(iv) Comment on the value of rate constant when 60 or 70% aqueous ethyl alcohol is used.

(v) Compare these values with that obtained when 80% aqueous ethyl alcohol was used. Account for any differences observed.

EXPERIMENT 26
Determination of Order for the Persulphate - Iodide Reaction

Aim: To determine the individual orders with respect to iodide and persulphate and the total order of the reaction of oxidation of iodide ion by persulphate.

Apparatus Required: Burette 50 cm^3, pipettes 5, 25, and 50 cm^3, stoppered bottles, 250 cm^3 conical flasks, 250 cm^3 standard flasks, porcelain trough, porcelain tiles.

Chemicals Required: Potassium iodide solution (0.1 N); Potassium persulphate solution (0.1 N); Acetic acid (1.0 N); Sodium thiosulphate solution (0.01 N); Starch indicator, ice cold water.

Principle: The overall oxidation of iodide ion by persulphate can be expressed as

$$2\,I^- + S_2O_8^{2-} \longrightarrow 2\,SO_4^{2-} + I_2$$

or more explicitly

$$S_2O_8^{2-} + I^- \longrightarrow SO_4^{2-} + SO_4^- + I^\circ$$

$$SO_4^- + I^- \longrightarrow SO_4^{2-} + I^\circ$$

$$I^\circ + I^\circ \longrightarrow I_2$$

The rate of the reaction is followed by estimating the iodine formed at different time intervals, by titrating with sodium thiosulphate using starch as indicator. The volume of thiosulphate is plotted as a function of time. The initial slope of this plot gives the initial rate of the reaction. The values of initial rates obtained can be used to calculate the total order and individual orders with respect to iodide as well as persulphate ion.

Procedure: Different reaction mixtures using the volumes given in the Table 1 are prepared. For example, mix 5 cm^3 of 1.0 N acetic acid and 100 cm^3 of 0.1 N potassium iodide in a stoppered bottle. Thermostat the same. Add the required amount of water (in this case 45 cm^3) so that the final volume is 200 cm^3. Pipette out 50 cm^3 of 0.1 N potassium persulphate solution in the stoppered bottle, starting a stopwatch when half the volume of the persulphate is added to the bottle. Completely stir the reaction mixture and pipette out 5 cm^3 of the reaction mixture into a conical flask containing ice cold water to quench the reaction and titrate the iodine liberated against thiosulphate solution using starch as the indicator. Repeat this at every 5 minutes interval for at least 40 minutes. Carry out similar titrations with the other combination of solution mixtures in the table below. Standardise the thiosulphate solution.

Table 1. Typical reaction mixtures for determining the order of the reaction between iodide and persulphate ions.

Bottle no.	Volume of acetic acid (1.0 N) (cm^3)	Volume of 0.1 N KI solution (cm^3)	Volume of 0.1 N $K_2S_2O_8$ solution (cm^3)	Volume of water in (cm^3)
1	5	100	50	45
2	5	50	25	120
3	5	100	25	70
4	5	50	50	95

Treatment of Data: Record all your observations systematically as follows:

Construct separate tables for each reaction mixture as given in Table 2. Plot the titre value as a function of time and evaluate the initial slopes (this can be done either numerically, or graphically or by both procedures). The slopes obtained at the initial periods of the reaction can be taken to be initial rates. (Reason out why we are interested in the initial rates and not the values of rates at any given time of the reaction).

Table. 2. Volume of thiosulphate consumed for a known aliquot of the reaction mixture at various time intervals.

Normality of thiosulphate solution taken = . . . N

Reaction mixture (bottle no.1)

Time in minutes	Burette readings		Volume of thiosulphate (cm^3)
	Initial	Final	
0			
5			
10			
15			
20			
25			
30			

Normally in a chemical reaction the rate of the reaction is proportional to the concentration of the reactants raised to the power m, where m is the order of the reaction with respect to that reactant. For the reaction between iodide and persulphate ions the rate expression can be written as

$$\text{Rate } \alpha \ [I^-]^m \ [S_2O_8^{\ 2-}]^n = k \ [I^-]^m \ [S_2O_8^{\ 2-}]^n$$

The ratio of the initial rate values obtained for bottles 1 and 4 can be written as

$$\text{rate1/rate4} = k[100]^m \ [50]^n / k[50]^m \ [50]^n$$

(Here the volumes taken are assumed to be proportional to concentration since the total volume of the reaction mixtures is kept constant.)
Therefore,

$$\text{rate1/rate4} = [2]^m$$

or, $$\log \{\text{rate1/rate4}\} = m \log 2$$

So the value of m, the order with respect to iodide can be found out. Similarly the ratio of the rates for bottles 1 and 3 can be written as

$$\{\text{rate1/rate3}\} = k \ [100]^m \ [50]^n/k \ [100]^m \ [25]^n$$

or $$\log \{\text{rate1/rate3}\} = n \log 2$$

So the value of n, the order with respect to persulphate can be calculated. The overall order of the reaction = m + n.

Also, the ratio rate1/rate 2 = $k \ [100]^m \ [50]^n/k \ [50]^m \ [25]^n$

or, $\log \{\text{rate1/rate2}\} = (m + n) \log 2$

Hence, the overall order of the reaction (m + n) can be calculated.

Result:

Report the individual orders and overall order of the reaction. Comment how the observed reaction orders accounts for the mechanism of the reaction.

Note:

1. The reaction is also believed to occur in steps:

$$I^- + S_2O_8^{2-} \xrightarrow{\text{slow}} (S_2O_8I)^{3-}$$

$$(S_2O_8I)^{3-} + I^- \xrightarrow{\text{fast}} 2\,SO_4^{2-} + I_2$$

2. Suggest any alternate ways of studying the kinetics of the reaction between iodide and persulphate ions.

EXPERIMENT 27

Determination of Order for the Peroxide - Iodide Reaction

Aim: To evaluate the order with respect to the concentrations of the reactants in the oxidation of HI by H_2O_2.

Chemicals Required: Hydrogen peroxide solution (0.1 N), Hydrogen iodide solution (0.1 N), sodium thiosulphate solution (0.05 N), starch indicator.

Apparatus Required: Burette, pipettes, conical flasks.

Principle: The reaction proceeds in stages as per the following equations

$$HI + H_2O_2 \longrightarrow HIO + H_2O$$

$$HIO + HI \longrightarrow H_2O + I_2$$

The rate of the reaction can be followed by titrating the iodine liberated as a function of time with thiosulphate using starch as indicator.

Procedure: Prepare approximately 0.1 N solutions of HI and hydrogen peroxide. Pipette out 50 cm^3 of HI solution in a stoppered bottle. Thermostat the same. After temperature equilibrium, 25 cm^3 of hydrogen peroxide solution is added to the stoppered bottle. Start the stopwatch. Pipette out, as soon as possible, 5 cm^3 of the reaction mixture into a conical flask and titrate the contents with a 0.05 N solution of thiosulphate using starch as indicator.Repeat pipetting out 5 cm^3 aliquots at regular intervals and titrate against standard thiosulphate solution. Pipette out into another stoppered bottle 25 cm^3 of HI and 25 cm^3 of water and thermostat the same. Add 25 cm^3 of hydrogen peroxide solution and follow the rate of the reaction by titrating against thiosulphate solution, which has been standardized already. Plot the volume of the thiosulphate used for each titration against the time of the reaction at which the aliquots are pipetted out into the conical flask. Calculate the initial rates.

Observations: Volume of thiosulphate consumed for a known aliquot of the reaction mixture at different time intervals:

Normality of thiosulphate solution = N.

Reaction mixture — 1

S. No.	Time (s)	Burette readings		Volume of thio consumed (cm^3)
		Initial	Final	
1				
2				
3				
4				
5				
6				

Construct similar tables for other reaction mixtures taken.

Treatment of Results: The initial rates are arrived at from the plots of concentration versus time. Using the initial rates, the order can be evaluated as shown below.

$$\text{Rate} = k\ [HI]^n\ [II_2O_2]^m$$

$$\text{Rate 1} = k\ [HI]_1^{\ n}\ [H_2O_2]_1^{\ m}$$

$$\text{Rate 2} = k\ [HI]_2^{\ n}\ [H_2O_2]_2^{\ m}$$

Between these two experiments the concentration of hydrogen peroxide has remained constant while the ratio of the concentrations of hydrogen iodide is 2:1. Therefore

$$\frac{\text{rate 1}}{\text{rate 2}} = \frac{[HI]_1\ [H_2O_2]_1}{[HI]_2\ [H_2O_2]_2} \quad \text{or} \quad \frac{\text{rate 1}}{\text{rate 2}} = 2^n$$

Hence, the value of n can be calculated.

Similarly another set of experiments were carried out keeping the concentration of HI constant by varying the concentration of hydrogen peroxide by a factor of 2. This will enable the evaluation of the value of m, the order with respect to H_2O_2.

Result:

Report the order with respect to HI and hydrogen peroxide and confirm the mechanism proposed.

EXPERIMENT 28
Kinetics of Iodination of Acetone by Spectrophotometry

Aim: To study the kinetics of iodination of acetone spectrophotometrically.

Chemicals Required: Iodine solution –0.005 M, hydrochloric acid – 0.5 M, acetone – 0.5 M (each about 100 cm^3).

Apparatus Required: Standard flasks, spectrophotometer and cells, and pipettes.

Principle: The rate law for any reaction can be obtained only by experimentally determining the species whose concentration affects the rate of the reaction and knowing the functional dependence of this concentration term with respect to the observed rate. The overall stoichiometric equation of the iodination of acetone is given by

$$CH_3 - CO - CH_3 + I_2 \longrightarrow CH_3COCH_2I + HI$$

The reaction rate is found to depend on the concentration of the acid used. The overall reaction sequence can be written as

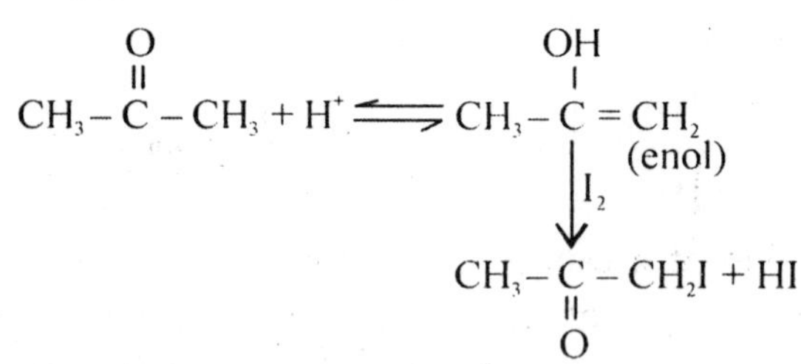

The order of the overall reaction as well as order with respect to individual components namely acetone, $[H^+]$ and iodine can be determined by measuring the rate of the reaction at varying concentrations of the reagents. A set of reaction mixtures as given in Table.1 is prepared at the appropriate time and the absorbance of the reaction mixture was measured for each of these solutions at 450 nm.

Procedure: Prepare the six different reaction mixtures as given in Table 1. After preparing each reaction mixture (always add iodine solution finally), take the well-mixed reaction mixture (about 2 – 4 mL) in the spectrometer cell and record the values of absorbance at 450 nm, as a function of time, preferably at 2 minute interval till the absorption value decreases to at least to 25 % of its initial value.

Table 1. Set of reaction mixtures to be prepared:

Volume of acetone (cm^3)	Volume of HCl (cm^3)	Volume of iodine (cm^3)	Volume of water (cm^3)
1.0	5.0	1.0	3.0
1.0	5.0	2.0	2.0
1.0	5.0	3.0	1.0
0.5	5.0	2.0	2.5
2.0	5.0	2.0	1.0
2.0	2.5	2.0	3.5

Treatment of Data: Record the absorbance as a function of time in a tabular form as given below in Table 2.

Table 2.

Time (min)	
Absorbance	

Construct similar tables for all the set of experiments done.

Plot absorbance data as a function of time for each of the reaction mixture and calculate the rate values. The values of rates obtained for reaction mixture 1 and 2 leads to the expression, since the reaction rate is proportional to concentrations,

$$\text{rate} = [I_2]^a\ [\text{acetone}]^b\ [H^+]^c$$

or, $\text{rate}_1/\text{rate}_2 = k\,[I_2]_1{}^a/k\,[I_2]_2{}^a$ (where the subscripts denote the reaction number).

Since concentrations of acetone and H^+ are the same for these two reaction mixtures and the concentration of iodine is in the ratio of 2 : 1, one can write,

$$\text{rate}_2/\text{rate}_1 = 2^a$$

or, $\log \{rate_2/rate_1\} = a \log 2$

or $a = \log \{rate_2/rate_1\}/\log 2$

Similarly, from the rate ratios of the appropriate reactions where the concentrations of acetone and acid are different, the values of b and c respectively can be calculated.

The values of a, b, c, are the order with respect to iodine, acetone and H^+ respectively.

Result:

Report the order on acetone, iodine and acid.

Note:

(i) The reported order of the reaction with respect to each acetone and acid is one and with respect to iodine is zero.

(ii) Rationalize this observation in terms of the mechanism of the reaction.

EXPERIMENT 29
Evaluation of First Order Rate Constant by Potentiometry

Aim: To determine the value of the pseudo first order rate constant for the oxidation of formic acid by bromine potentiometrically.

Apparatus Required: Calomel electrode, sensitive potential measuring device, standard flasks and pipettes.

Chemicals Required: Bromine 0.03 M, formic acid 0.40 M, HCl 2 M, NaBr 1 M.

These solutions may be diluted accordingly as required in the procedure.

Principle: Set up a galvanic cell of the type

$$^{-}Hg, Hg_2Cl_2 \mid Cl^- \parallel Br^-, \; Br_2 \mid Pt^{+}$$

The platinum electrode needs cleaning which is achieved by heating the wire to red heat for a minute at the beginning of each day of the experiment. The EMF of the cell is given by

$$E = E°Br_{2,\,Br-} + (RT/2F) \ln \{[Br_2]/[Br^-]\} - E_{cal}$$

If the bromide concentration is very large so that it changes negligibly during the experiment one can write

$$E = \text{constant} + (RT/2F) \ln [Br_2]$$

If the reaction is pseudo first order with respect to Br_2 then one can write $\ln [Br_2]$ = constant – k_p^t, where k_p is the pseudo first order rate constant.

Since cell EMF will be a linear function of time, from the slope of this linear plot one can calculate the value of k_p which is given by

$$k_p = -\left(\frac{2F}{RT}\right)\left(\frac{dE}{dt}\right).$$

The order with respect to formic acid, H^+ and Br^- can also be determined by halving and/or doubling these concentrations one by one and repeating the experiment.

Procedure: Prepare in suitable volumetric flasks formic acid, and HCl mixture such that concentrations with respect to each of them is 0.1 M. Pour this into the reaction cell with arrangement for stirring (see Fig. 18). Prepare bromine – bromide mixture in another standard flask such that $[Br_2]$ is 0.003 M and $[Br^-]$ is 0.1 M. Equilibrate both these solutions in a water bath separately. Pour the bromide/bromine mixture into the reaction cell and start a stop watch. Couple the reaction cell to a calomel electrode and connect the calomel electrode to the negative end of the potentiometer and the Pt electrode to the positive end of the potentiometer. At various time intervals say 1, 2, 4, 5, 8, 10 minutes adjust the potentiometer and read the potential.

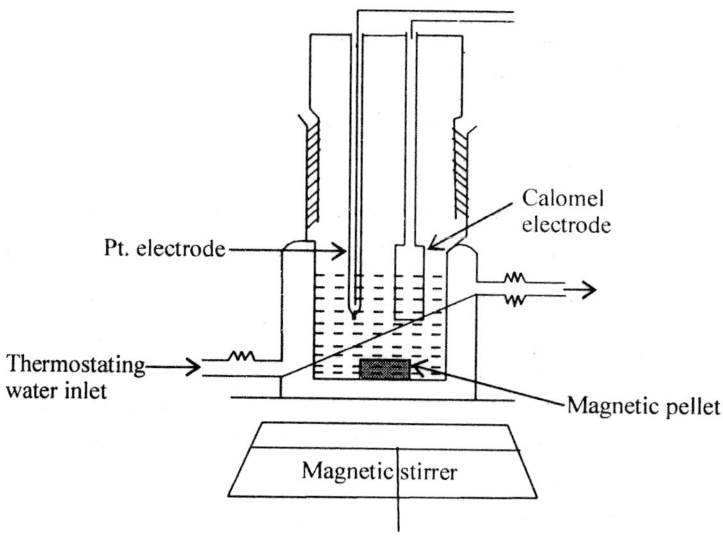

Fig. 18. Reaction assembly for potentiometric study of reaction kinetics.

Treatment of Data: Tabulate the data as shown below :

[HCOOH] = M; $[Br_2]$ = M [HCl] = M; Temperature = °C

Time (min)	
Potential (mV)	

1. Plot EMF versus time. The linearity of this plot shows that the reaction is first order with respect to bromine concentration.

2. Evaluate the slope of this plot and calculate the value of pseudo first order rate constant, k_p which is given as

$$k_p = -\left(\frac{2F}{RT}\right) \times \text{slope.}$$

Result:

Report the value of the pseudo-first order rate constant for the oxidation of formic acid by bromine.

Note:

At high concentrations of bromide, one has to take into account the tribromide formation equilibrium, $Br_2 + Br^- \rightleftharpoons Br_3^-$ so that the total analytical concentration of bromine ($[Br_2] + [Br_3^-]$) in acid solution is given by

$$[Br_2]_a = [Br_2]\,[1 + K_t\,[Br^-]],$$

where K_t is the formation constant of tribromide. In this case, the rate of the reaction is given by $-\,d[Br_2]_a/dt$ which is equal to $-\,d\,[HCOOH]/dt$ or $d[CO_2]/dt$.

EXPERIMENT 30

Kinetics of a Polymerization Reaction

Aim: To study the kinetics of polymerization of acrylamide initiated by Ce^{4+} and malonic acid.

Chemicals Required: Ce^{4+} solution 0.05 M; malonic acid (MA) 0.5 M; acrylamide 1 M; sulphuric acid 10 N ; bromate – bromide mixture, 10 % KI solution, sodium thiosulphate solution, ferrous sulphate 0.02 M; nitrogen gas.

Principle: The sequence of chemical reactions can be represented as

$$Ce^{4+} + MA \xrightarrow{K_r} Ce^{3+} + \underset{(R)}{CH(COOH)_2} + H^+$$

$$R + Ce^{4+} \xrightarrow{k_0} \text{product}$$

$$R + \underset{(M)}{CH_2{=}CH-CONH_2} \xrightarrow{k_1} \underset{[R\text{-}M]}{R-CH_2-\cdot CH-CONH_2}$$

This chain may progress with further reactions with R – M˙ with M. forming the polymers $R-(M)_n$

The termination step can be $R-M\cdot + R-M\cdot \xrightarrow{k_t}$ dimer

Therefore, applying steady state principles for R• we get

$$\frac{[dR\cdot]}{dt} = k_r [Ce^{4+}] [MA] - k_o [R\cdot] [Ce^{4+}] - k_i [R\cdot] [M] = 0$$

$$[R\cdot] = k_r [Ce^{4+}] [MA]/\{k_o [Ce^{4+}] + k_i [M]\}$$

Similarly, applying the steady state principles to other radicals in the propagation steps and simplifying the results, ultimately it can be shown that

$$\text{Rate of polymerization} = k_p [R-(M)_n] [M]$$

$$= \frac{k_p k_i k_r [M]^2 [MA] [Ce^{4+}]}{k_t [Ce^{4+}] \{k_0 [Ce^{4+}] + k_1 [M]\}} \qquad (1)$$

If k_1 [M] << k_o [Ce^{4+}], then the rate law reduces to

$$\text{Rate} = \frac{k_p k_i k_r [M]^2 [MA]}{k_0 k_t [Ce^{4+}]} = k_{obs} \frac{[M]^2 [MA]}{[Ce^{4+}]} \quad (2)$$

On the other hand if k_1 [M] >> k_o [Ce^{4+}]

$$\text{Rate} = \frac{k_p k_i k_r [M]^2 [MA]}{k_t k_i [M]} = \frac{k_p k_r}{k_t}[M][MA] = k_{obs} [M][MA] \quad (3)$$

If k_0 [Ce^{4+}] ≈ k_1 [M] then the rate is given by expression (1)

Procedure: Prepare different reaction mixtures keeping the concentration of all the constituents constant and varying the concentration of monomer only. Typical concentrations can be

[Ce^{4+}] = 0.005 M ; [Acrylamide] = 0.2 – 0.8 M (variable)

[H_2SO_4] = 0.5 N [MA] = 0.05 M.

Keep the total volume of the reaction mixture constant. The solutions should be made free of oxygen by passing nitrogen gas. Thermostat the solutions and add malonic acid as the last constituent and start the stopwatch. At various time intervals, the amount of unreacted monomer can be estimated by bromometry after arresting the reaction by addition of ferrous sulphate solution in 1 N sulphuric acid. Estimate the original concentration of the monomer used also by bromometry. For a known volume of the monomer, add 2 cm^3 of bromate/bromide mixture and 10 cm^3 of 2 N sulphuric acid. Keep it in the dark for 15 minutes in an iodine flask, add KI solution and titrate the liberated iodine against thiosulphate solution of known strength. Also estimate the thiosulphate required for titrating the known volume of bromate/bromide mixture.

Treatment of Data: Fill in the titre values in the following table:

Time (min)								
0.2 M acrylamide								
0.4 M acrylamide								
0.6 M acrylamide								
0.8 M acrylamide								

From the titre values at various time intervals (or for a fixed time interval) one can get the value of the rate of polymerization, that is – d[M]/dt.

Then, plot the logarithm of the rate versus log [M] and get the value of the slope. If the slope is 2, then the condition $k_o [Ce^{4+}] >> k_l[M]$ is satisfied.

If the value of the slope is 1 then the condition $k_l [M] >> k_o [Ce^{4+}]$ holds good.

Then, by substituting the appropriate concentrations in equations (2) or (3), the rate constant k_{obs}, can be evaluated.

If the value of the slope is fractional then neither $k_l[M]$ nor $k_o[Ce^{4+}]$ can be neglected in comparison with the other.

(Note: If slope is 1.5 then the reaction can be accounted for in terms of another mechanism).

Result:

Comment on the order of the reaction with respect to the monomer. Outline the actual mechanism of the reaction.

EXPERIMENT 31
Kinetics of Catalytic Decomposition of Hydrogen Peroxide

Aim: To study the kinetics of the catalytic decomposition of hydrogen peroxide by manganese (IV) oxide.

Chemicals Required: Hydrogen peroxide 20 volume % solution, Manganese (IV) oxide – free of any carboneous impurity.

Apparatus Required: The apparatus has to be assembled as shown in Fig. 19.

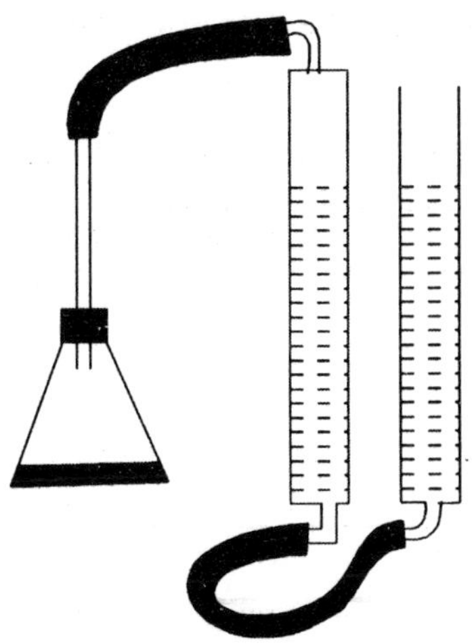

Fig. 19. Typical experimental arrangement for the study of catalytic decomposition of hydrogen peroxide solution.

Principle: The overall reaction of the decomposotion of H_2O_2 can be written as

$$2H_2O_2 \longrightarrow 2H_2O + O_2$$

The reaction can be described either by a first order rate equation or by a second order rate equation. The analysis of data has to be carried out for both these orders to decide which rate equation is applicable for the kinetics of the catalyzed decomposition of hydrogen peroxide.

Procedure: Take 90 cm^3 of distilled water and 10 cm^3 of 20 volume hydrogen peroxide in a conical flask fitted with one hole stopper with a suitable arrangement for the collection of gas as well as a side tube to adjust the pressure in the collection tube (Refer the Fig. 19). Add the catalyst, 0.1 g of powdered manganese (IV) oxide to the reaction vessel in a capsule and then close it, simultaneously starting a stopwatch. [In the case of a closed capsule, the stopwatch is started when the capsule is broken in situ by means of a suitable magnetic pellet stirrer.] The conical flask containing the reactant is to be thermostated before the addition of manganese (IV) oxide. The volume of oxygen evolved is measured from the gas burette after adjusting the liquid levels equal, both in the gas collection burette as well as in the side limb. Record the volume of the gas evolved as a function of time.

Treatment of Data: Tabulate the data of the volume of the gas collected as a function of time.

Table 1: Data of the volume of gas (oxygen) collected as a function of time.

Time (s)	0	30	60	90	120	150	180	210	240	270	300	330	360	390	420
Volume of gas collected at NTP (cm^3)															

If V is the total volume of oxygen that can be obtained from the initial volume of hydrogen peroxide taken and v is the volume of oxygen evolved at any given time, then the rate of the reaction is given by

$$dv/dt = k\,(V - v)^n$$

where k is the specific reaction rate and n is the order of the reaction. If n were to be 1, then a plot of logarithm of $V/(V - v)$ versus time should be a straight line passing through the origin.

If n were to be 2, then a plot of $\{(1/(V - v)) - (1/v)\}$ versus time should be a straight line. Plot both sets of data on the same graph paper. Find out which data gives a better straight line. (You may use a linear regression fit and compare the correlation coefficient values).

Results:

1. Report the kinetic order of the reaction.
2. Justify the mechanism of the reaction.
3. Comment on the sources of error as well as the reasons for the deviations in the plots in the initial time periods.

Note:

The solubility of oxygen in water would vitiate the results in the initial periods of the reaction. Suitable correction is possible, if the solubility of oxygen in water at the experimental temperature is known and the quantity of water present in the reaction vessel can be correctly estimated as a function of time. Secondly, since the reaction is exothermic, the heat evolved may alter the temperature of the reaction. Either vigorous stirring or properly thermostating the reaction solutions may yield better results.

EXPERIMENT 32
Thermal Decomposition of Potassium Chlorate

Aim: To study the thermal decomposition of potassium chlorate in presence of manganese (IV) oxide.

Chemicals Required: Potassium chlorate, Manganese (IV) oxide–free from carboneous impurities.

Apparatus Required: The apparatus has to be assembled as shown in Fig. 20.

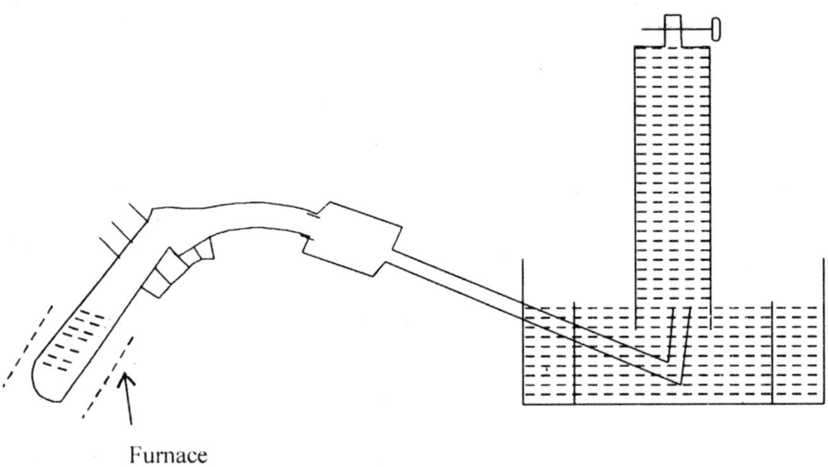

Fig. 20. Typical experimental arrangement for the study of decomposition of $KClO_3$

Procedure: Take 5 g of potassium chlorate in a dry hard glass (boiling) tube with a one-holed stopper connected to a delivery tube so that the gas evolved can be collected by the downward displacement of water (Fig. 20). Heat the boiling tube containing the reaction mixture by means of an electric heater at constant temperature. Collect the gas evolved by noting the readings in the inverted gas burette. Now replace the 5 g $KClO_3$ with a mixture of 5 g of potassium chlorate and 0.1 g of Manganese (IV) oxide and heat it at the same temperature using the same heater. Note the volume of gas evolved as a function of time. Repeat the experiment with different amounts of manganese (IV) oxide in the range 0.01 – 1.0 g and tabulate the volume of gas evolved as a function of time for different quantities of manganese (IV) oxide used as catalyst.

Treatment of Data: Tabulate the volume of oxygen evolved as a function of time .

$KClO_3$: 5 g Laboratory temperature: . . . °C

Time (min)								
0.05 g MnO_2								
0.10 g MnO_2								
0.2 g MnO_2								
0.5 g MnO_2								

1. Plot the volume of the gas evolved as a function of time for various amounts of catalyst added.
2. Tabulate the data on the weight of the catalyst against the time taken for the total volume of gas to be evolved.
3. Attempt to generate suitable rate data (even reciprocal of time for complete conversion can be taken as a measure of this quantity) as a function of the weight of the catalyst. Plot this data which will reflect how the presence of catalyst affects the rate of the reaction.

Results:

1. Report on how increasing MnO_2 mass affect the rate.
2. Comment on the (i) possible sources of experimental error. (ii) solubility of oxygen in water and how does it affect the rate data generated, and (iii) mechanism of the thermal decomposition of potassium chlorate.

Note:

The following precautions are necessary for this experiment.

(i) It is desirable to prepare and label all the mixtures at the start of the experiment itself.

(ii) The gas burette has to be specially made for this purpose to hold gas volumes upto 2000 cm^3. The gas tube should be carefully calibrated before use. It is possible to use a suitable diameter glass tube with a stop cork attached at the top and a graph paper pasted over it as the gas burette. The calibration marks can be indicated in the graph paper pasted on the tube.

(iii) It is necessary that the quantity of manganese (IV) oxide should not exceed 0.8 to 1.0 g for 5 g of potassium chlorate. Suitable protection in the form of perspex shielding screen can be used while carrying out the thermal decomposition reaction.

(iv) Also, before the start of the experiment, calculate the amount of gas that will be evolved at NTP for the complete decomposition of potassium chlorate taken, at the laboratory temperature. (theoretical yield for 5 g of potassium chlorate will be around 1380 cm^3 at NTP).

(v) The raw data of volume of gas evolved as a function of time can also be analysed in terms of various types of equations applicable for the thermal decomposition of solids. For this, go through the text dealing with the solid state decomposition reactions and note down the equations applicable to such decomposition. The data have to be reduced in terms of fraction (α) of potassium chlorate decomposed as a function of time.

EXPERIMENT 33
Effect of Surface Area of Catalyst and Temperature on the Kinetics of Metal-Acid Reaction

Aim: To study the effects of concentration, temperature and surface area of catalysts on the reaction between a metal and an acid.

Chemicals Required: Magnesium wire, magnesium ribbon and granular zinc; 0.5 M and 2.0 M Hydrochloric acid.

Apparatus Required: The apparatus has to be assembled as shown in Fig. 21.

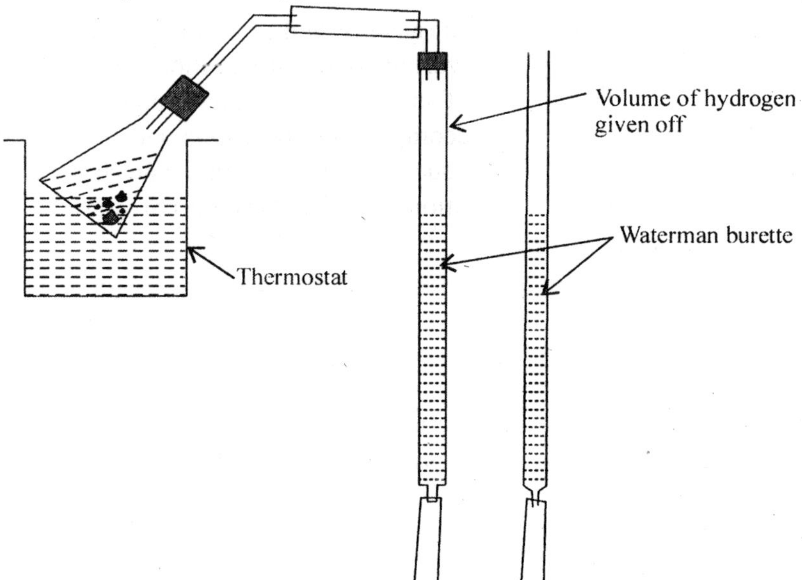

Fig. 21. Typical experimental arrangement for the study of kinetics of hydrogen evolution from an acid by a metal

Procedure: Place a piece of magnesium wire (0.05 g) by sliding it into the clamped conical flask containing 100 cm^3 of HCl. Close the apparatus. Adjust the water levels in the gas burette to that of the adjacent limb which is exposed to atmosphere. By doing this adjustment, the pressure of the gas in the burette becomes equal to the atmospheric pressure. All gas volume readings are recorded after adjusting the water levels in the burette and the adjacent limb exposed to atmosphere to be the same. After magnesium wire is slid into the flask, start a stopwatch. Record the volumes of gas collected every 30-60 seconds at regular time intervals till 40-50 cm^3 have been evolved or a time period of 10-12 minutes has elapsed after the start of the reaction. Tabulate the volume of gas evolved as a function of temperature of the reaction also. (This experiment can also be repeated with a weak acid like acetic acid).

Carry out the same experiment with 100 cm^3 of 0.5 M HCl with 0.05 g of magnesium ribbon which will have higher surface area compared to the same weight of magnesium wire.

Repeat the experiment using 100 cm^3 of 2 M hydrochloric acid and 0.3 g of zinc at steady temperatures of 30, 35 and 40^{0}C.

Treatment of Data: Tabulate the volume of gas evolved as a function of time in each case appropriately.

Plot the volume of gas collected as a function of time in every set of experiments done.

Compare the data obtained with 0.05 g of magnesium wire and ribbon.

Deduce the effect of surface area of the magnesium material used on the rate of the reaction.

Plot the logarithm of the rate of hydrogen production versus reciprocal of the absolute temperature and deduce the value of the activation energy of the reaction.

Result:

Comment on the effect of surface area and nature of the metallic material used on the rate of the reaction. Report the value of the activation energy of the reaction.

Note:

The weight of the magnesium wire and the ribbon used in the experiment should be the same, only then the effect of surface area on the reaction rate can be compared.

EXPERIMENT 34
Catalytic Constant of an Acid

Aim: To find out the catalytic constant of hydrochloric acid with respect to the iodination of acetone.

Chemicals Required: Acetone, iodine (0.05 M in 10% KI), thio (0.01 M) hydrochloric acid (1 M), sodium acetate solution (10%), starch indicator.

Principle: When the rate of any reaction increases with increase in the added acid concentration, it is said to be an acid—catalysed reaction. For such reactions, the experimentally observed rate constant will be given as

$$k_{obs} = k_o + k_H + [H^+]$$

where k_o and k_H^+ are the respective values of the rate constants in the absence and presence of the acid catalyst. The coefficient k_H^+ is known as the catalytic constant of the acid and is a measure of the contribution of the added H^+ to the rate increment.

Procedure: Place 10 cm^3 of acetone and 20 cm^3 of 1 M HCl in a 250 cm^3 standard flask and dilute it to the mark. Place the contents in a 500 cm^3 conical flask and immerse this in a water bath at 298 K until thermal equilibrium is attained. Pipette out 20 cm^3 of the iodine solution (previously warmed to the bath temperature) into the flask. vigorously mix the contents and start the stop clock simultaneously. Pipette out 20 cm^3 of the reaction mixture as soon as possible after starting the reaction into 10 cm^3 of 10% sodium acetate solution and titrate the residual iodine with thio to a starch end point. Repeat this at every 5 minutes intervals.

Treatment of Data : Tabulate the data obtained as follows:

Time (min)	
Vol. of thio (cm^3)	

Using the density of acetone, calculate its initial concentration in the reaction mixture From the value of the initial concentration of iodine obtained from the first titration (V_0), and the concentration of iodine remaining at any other interval (V_t), the iodine reacted ($V_0 - V_t$), at that time can be calculated. (This is also equal to the amount of acetone reacted and from the difference between the initial acetone concentration and that reacted at time t, the acetone concentration remaining at time t is calculated). Plot the values of $\log(V_0 - V_t)$ with time and from the slope of the straight line obtained find the rate constant for the reaction.

Repeat the experiment for 5, 10 and 25 cm^3 of 1 M HCl keeping the [acetone] constant. Make a plot of the rate constant versus $[H^+]$ and from the slope, calculate the catalytic constant of H^+ ion.

Result:

Report the values of rate constants evaluated at different [HCl] and the value of the catalytic constant for HCl.

EXPERIMENT 35
Mechanism of the Oxidation of an Organic Compound from Kinetic Data

Aim: To study the kinetics of oxidation of an organic substrate and elucidate the mechanism of the reaction.

Apparatus Required: Burettes, pipettes, conical flasks, thermostat, bottles, stopwatch.

Chemicals Required: Stock solution of oxidant (Ce^{4+} solution in 1-2 N sulphuric acid, 2 N sulphuric acid, standard ferrous ammonium sulphate solution, ferroin indicator, suitable organic substrates, (distilled) phosphoric acid.

Principle: The kinetics is followed under pseudo first order conditions which involves using excess substrate and monitoring oxidant concentration as a function of time. These pseudo first order plots of log [oxidant] versus time yield data on $k_{observed}$. The variation in $k_{observed}$ as a function of substrate concentration or acidity can give some idea on the mechanism of the reaction. If a spectrophotometer is used to record the absorption due to the oxidant only versus time and the data can be treated in a similar manner. However in this case extremely low concentrations have to be used.

Procedure: Prepare the different reaction mixtures by diluting appropriately the stock solutions so that each of the reagents will have the respective concentration as given in the following table. The total volume is made upto 50 cm^3. The oxidant stock is estimated using the ferrous ammonium sulphate solution before the start of the experiment. Add all the reagents but the oxidant and thermostat the bottles in a trough for room temperature (or in bath for higher temperatures). Thermostat the oxidant separately and when the temperature equilibrium is attained, transfer the appropriate volume of oxidant into the other flask and mix the solutions. While running down the oxidant into

into the other flask and mix the solutions. While running down the oxidant into the other flask, at mid way start the stop watch. At 10-15 minute intervals remove 5-10 cm^3 of the reaction mixture and add the same to 10-20 cm^3 ferrous ammonium sulphate taken in a conical flask. Add 5 cm^3 of phosphoric acid and titrate excess ferrous ammonium sulphate against standard ceric solution in the burette with ferroin as indicator. Calculate the oxidant remaining at time t.

Composition of Reaction Mixtures

[substrate] M	$[Ce^{4+}]$	[acid] M
0.2	0.02	1.0
0.4	0.02	1.0
0.6	0.02	1.0
0.4	0.01	1.0
0.4	0.005	1.0
0.4	0.025	1.0
0.4	0.02	1.5
0.4	0.02	0.5

Treatment of Data: Tabulate the titre values (or the absorbances) noted at different time intervals for each of the reactions.

1. Plot log [oxidant] versus time for each reaction. From the slope get the value of $k_{observed}$.
2. Plot log[substrate] versus log k for reactions with different substrate concentrations but with the same initial concentration of oxidant. The slope of this plot will give the order on substrate, if it is linear.
3. Plot the log $k_{observed}$ versus $\log[H^+]$; if it is linear, the slope gives the order on the acid.
4. Plot $k_{observed}$ versus [substrate]. If the plot is linear, calculate $k_{second\ order}$. If not try plotting $(k_{observed})^{-1}$ versus $[substrate]^{-1}$. If this plot is linear obtain the slope and intercept of the plot. These lead to data on the first order rate constant (k) and the equilibrium constant (K) for the following mechanism.

$$Ox + sub \underset{}{\overset{k}{\rightleftharpoons}} complex$$

complex ⟶ product

This mechanism would lead to Michaelis-Menton type kinetics, with k being first order rate constant.

Result:

Report the rate constants at different compositions of reaction mixtures.

Report the order on the substrate, oxidant and acid.

Suggest a suitable mechanism for the reaction.

Note:

1. The substrate to be used in this experiment may preferably be a primary or a secondary alcohol.
2. In all such reactions the stoichiometry of oxidant consumed per mole of substrate is generally larger than one, which means that several of the initial products are capable of being oxidized further. Provided these subsequent reactions are all rapid compared to the initial reaction and provided one confines oneself to really initial rates, the data may be reliable.

EXPERIMENT 36
Rate Constant for the Decomposition of Hydrogen Peroxide Over Platinum

Aim: To determine the rate constant for the decomposition of hydrogen peroxide catalyzed by platinum.

Chemicals Required: Hydrogen peroxide, platinum catalyst (powder), N/100 potassium permanganate solution, 2 N H_2SO_4 solution.

Apparatus Required: Conical flask, standard flask, thermostat, burette, 10 cm^3 pipette.

Principle: The decomposition of hydrogen peroxide in aqueous medium is represented as

$$H_2O_2 \longrightarrow H_2 + \frac{1}{2}O_2$$

This reaction is normally catalyzed by colloidal platinum. The rate of the reaction can be followed either by monitoring the amount of gaseous oxygen evolved or estimating the amount of unreacted hydrogen peroxide by titrating against standard $KMnO_4$ in acid medium at different intervals of time.

$$5\,H_2O_2 + 2\,KMnO_4 + 3\,H_2SO_4 \rightarrow 2\,K_2SO_4 + 2\,MnSO_4 + 8\,H_2O + 5\,O_2\uparrow$$

Procedure: Take about 100 cm^3 N/100 H_2O_2 solution in a conical flask and dilute it with distilled water, if necessary, and thermostat it. Add the platinum catalyst to the conical flask. Shake the mixture vigorously and withdraw 10 cm^3 portions at every 2 minutes in the initial stages and at longer periods of interval (say 5 and 10 min) after about 15 minutes from the start of the reaction. Pipette out 10 cm^3 portions of the reaction mixture into another conical flask containing excess of dilute (2 N) sulphuric acid. This will arrest the progress of the reaction. Titrate this mixture against N/100 $KMnO_4$ taken in the burette.

Treatment of Data: Tabulate the titre value obtained at different time intervals. From the titre value, calculate the values of the first order rate constant using the expression

$$k = (2.303/t)\ \log\{a/(a - x)\}$$

The titre value at zero time = ...cm^3 (a)   Temperature = °C.

Time (min)	Titre value in cm^3 = (a – x)	k min^{-1}
2		
4		
6		
8		
10		
15		
20		
25		
30		
40		
50		
60		
80		
100		

Mean value k = ...min^{-1}.

Also, evaluate the value of k graphically from the slope of the log(a-x) versus time plot.

Result:

Report the values of the first order rate constant obtained by calculation and from the graph.

Note:

1. As the ratio of concentrations rather than the absolute concentration are required in the rate constant equation, the titre values themselves can be used to calculate the rate constant. The volume of $KMnO_4$ consumed at zero time of the titration, for titrating the first aliquot of the reaction mixture will be proportional to the initial concentration 'a' of H_2O_2 taken. The titre values at any time't' is proportional to the concentration of H_2O_2 remaining at time't', hence they can be taken to be (a-x).
2. Also observe the rate of decomposition of H_2O_2 in the absence of the catalyst and comment.

EXPERIMENT 37

Polymerisation of Methylmethacrylate

Aim: To study the kinetics of polymerization of methylmethacrylate

Chemicals Required: Benzoyl peroxide, methylmethacrylate. Benzene, methanol.

Apparatus Required: Good balance, conical flasks, stop watch.

Principle: Methylmethacrylate polymerization is initiated by peroxide.

$$CH_2 = CH - COOCH_3 \xrightarrow{\text{peroxide}} \text{polymer}$$

The polymer being insoluble in methanol, it is precipitated in methanol medium at different time intervals and weighed. Thus the progress of the polymerization is followed gravimetrically and from the weight, the rate constant is evaluated.

Procedure: Take a known volume (about 2.0 mL) of methylmethacrylate in pure benzene (about 20 mL). Add known amount of peroxide (say, 0.1 g) and start a stop watch. After 2 minutes pipette out known aliquots from the reaction mixture into a conical flask containing methanol. The polymer is precipitated which is filtered, dried and weighed. Repeat this for every 5 or 10 minutes interval for about an hour. Carry out the experiment at various concentrations of the monomer and that of the catalyst.

Treatment of Data: Note the experimental data in a table as shown below :

[Methylmethacrylate] = M; Catalyst = g; Temperature = ^{o}C.

Time (min)									
Weight (g)									

The rate of polymerization of methylmethacrylate is given as,

$$\text{rate} = k[M]^1[cat]^{1/2}.$$

Plot the weight of the polymer obtained as a function of time.

At various concentrations of monomer at constant concentrations of the catalyst the rate of polymerization must be a linear function of the concentration of monomer from which $k[cat]^{1/2}$ can be calculated. Knowing the value of the catalyst concentration, one can evaluate the rate constant of the polymerization reaction.

Result:

The rate constant for the polymerization of methylmethacrylate = ...

Note:

One can also determine the molecular weights of the polymer by determining the intrinsic viscosity as a function of monomer concentration and using the relationship $[\eta] = K[M]^a$, where M is monomer concentration.

EXPERIMENT 38
Light Intensity (In Einstein/Minute) Emitted by a Source – Actinometry

Aim: To determine the intensity of light in Einstein/min emitted from a source.

Chemicals required: 0.2% solution of 1, 10-phenanthroline, ferric sulphate solution, standardized 0.1 M EDTA solution, 0.08 M ferrous sulphate in 0.1 M sulphuric acid, 0.1 M potassium chromate, 1% diphenylamine in Con.H_2SO_4.

Apparatus Required: 25 cm^3 standard flasks, 100 cm^3 and 10 cm^3 volumetric flasks, 10 cm^3 graduated pipettes, spectrophotometer with cells.

Principle: The light energy emitted from any source is measured in the units of Einstein. One Einstein is equal to the energy absorbed or emitted by one mole of reactants. Light energy absorption is the primary process in any photochemical reaction. The photochemical reaction of interest here is

$$2Fe^{3+} + C_2O_4^{2-} \longrightarrow 2Fe^{2+} + 2CO_2$$

The amount of ferrous ions formed is determined as 1, 10–phenanthroline complex which absorbs at 510 nm.

Procedure: Prepare 0.08 M ferrous sulphate in 0.1 M sulphuric acid. Standardize the solution by titration against potassium chromate using diphenylamine as indicator. Now dilute the ferrous sulphate solution 200 times with 0.1 M sulphuric acid so as to obtain 4×10^{-4} M solution. In a series of 25 cm^3 volumetric flasks add 0, 1, 3, 5, 7, 9 cm^3 of the 4×10^{-4} M solution of ferrous sulphate. Add enough 0.1 M sulphuric acid to these 25 cm^3 standard flasks so that the volume is 12.5 cm^3. Add to each of these volumetric flasks 2 cm^3 of 0.2%, 1,10-phenanthroline solution. Add 6 cm^3 of the buffer solution to each of these flasks and make up to the mark.

Measure the absorbance of each of these solutions at 510 nm using 1 cm cell against the reference solution containing no ferrous sulphate i.e. the solution from the first volumetric flask. Plot the absorption values obtained against complex concentrations and determine the value of the extinction coefficient from the slope. (Alternatively a value of 1.11×10^4 can be used as the value of the extinction coefficient for the ferrous- phenanthroline complex.)

Prepare 0.2 N ferric sulphate solution (if necessary standardize against EDTA solution using salicyclic acid as indicator and a buffer with pH 3-4). Prepare a standard 1.2 M potassium oxalate solution. Pipette out 5 cm^3 each of the ferric sulphate solution and potassium oxalate solution into a 100 cm^3 volumetric flask. Make up to the mark. Pipette out 20-40 cm^3 of the $K_3Fe(C_2O_4)_3$ solution to a reaction vessel (may be a boiling tube) to be irradiated for a known period of time (say...min). Mix the irradiated solution and pipettc out 1 cm^3 of the actinometer solution into a 10 cm^3 volumetric flask. Add 2 cm^3 of 0.2% 1,10 phenanthroline and buffer solution. Now make up the solution and mix well. Prepare exactly similar solution with the non-irradiated sample using the same aliquot as the irradiated solution. Measure the absorbance of both the solutions at 510 nm.

Treatment of Data:

The volume of irradiated actinometer solution taken $= V_1$

The volume of actinometer solution originally irradiated $= V_2$

The volume to which the aliquot of the irradiated solutions is diluted

$= V_3 (10\ cm^3)$

Extinction coefficient of ferrous phenanthroline complex at 510 nm

$= \in (\sim 1.11 \times 10^4)$.

Quantum yield of Fe^{2+} production at wave length of light used

$= \phi\Lambda\ (\sim 1)$.

The time of irradiation $= T$

Using the absorption values, light intensity is calculated using the formula

$$I \text{ (in Einstein/min)} = AV_2V_3/\in\phi\wedge LTV_1$$

where A is the absorption of irradiated actinometer solution corrected for blanks.

Result:

Report the light intensity in Einstein/min.

EXPERIMENT 39
Inversion of Sucrose – Polarimetry

Aim: To determine the rate constant for the inversion of sucrose.

Apparatus Required: Polarimeter, volumetric flasks, pipettes and reaction bottles.

Chemicals Required: Pure sucrose, 1 N HCl and distilled water.

Principle: The reaction can be represented as

$$\underset{\text{Sucrose}}{C_{12}H_{22}O_{11}} + H_2O \longrightarrow \underset{\text{Glucose}}{C_6H_{12}O_6} + \underset{\text{Fructose}}{C_6H_{12}O_6}$$

The rate of this process is naturally dependent on the concentration of sugar, water and hydrogen ions. The concentration of water remains virtually constant (as water is used in excess) and as the hydrogen ions act as catalyst its concentration will be virtually remaining constant; thus, the reaction rate will be proportional to the concentration of sucrose only. Hence, the reaction becomes kinetically a first order reaction.

Sucrose solution is dextro-rotatory, but the mixture of glucose and fructose is leavo-rotatory. Therefore, the reaction can be followed by measuring the change in the angle of rotation by means of a polarimeter.

Procedure: Prepare 20% sucrose solution in pure distilled water. Set the polarimeter (Fig. 22) for the zero reading by filling distilled water in the polarimeter tube. Thermostat the sucrose solution and 1 N HCl solution in two bottles. When they have obtained thermal equilibrium, add 25 cm^3 of the thermostated acid to 25 cm^3 of the sugar solution and start a stop watch simultaneously taking care that the stop watch is started when half the volume is discharged into the sucrose solution. Mix the solution thoroughly and pour the reaction mixture in the polarimeter tube in such a way that there are no air bubbles in it. Determine the angle of rotation as soon as possible and mark the value as α_0. Record the angle of rotation at different time intervals for about 60 minutes and record each of them as α_t.

Finally, heat the reaction mixture at 80°C and use the solution after cooling to record the α_∞ value and tabulate the observations.

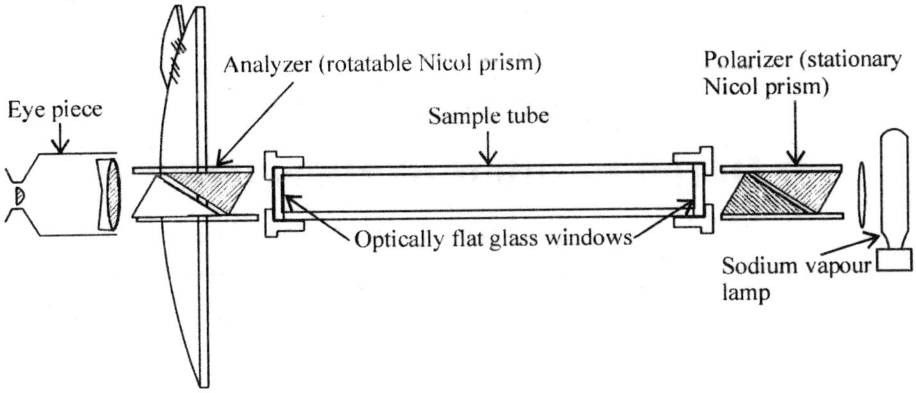

Fig. 22. Schematic diagram of a polarimeter.

Treatment of Data: Record the observed data in the following table.

α_0 = ... α_∞ = ...

Time (min)	Angle of rotation α_t	$\log(\alpha_t - \alpha_\infty)$	k
5			
10			
20			
30			
40			
50			
60			

Mean k = ...min^{-1}

Calculate the value of the first order rate constant using the various values of α_t in conjunction with α_0 and α_a employing the equation

$$k = (2.303/t) \log [(\alpha_0 - \alpha_\infty)/(\alpha_t - \alpha_\infty)]$$

Draw a graph of $\log(\alpha_t - \alpha_\infty)$ versus time and calculate the value of the rate constant from the slope of the straight line drawn.

Also calculate the half time required for the inversion of sucrose.

Result:

Report the value of the specific rate constant of the inversion of sucrose.

By calculation k = min^{-1}. From graph k = min^{-1}.

Half life of the reaction, $t_{1/2}$ = min.

EXPERIMENT 40
Decomposition of Diacetone Alcohol Using Dilatometer

Aim: To study the decomposition of diacetone alcohol by dilatometric method.

Chemicals Required: 0.1 M NaOH and diacetone alcohol.

Apparatus Required: Dilatometer, stop watch, burette and pipettes.

Principle: The reaction is represented as

$$(CH_3)_2-C(OH)-CH_2-CO-CH_3 \longrightarrow 2\,CH_3COCH_3$$

The reaction is catalyzed by hydroxyl ions and goes to completion in dilute aqueous solutions. Since there is a density difference between the reactant and the product, the reaction can be easily followed by the measurement of volume change as a function of time.

Guggenheim method (refer experiment 42) is adopted in this reaction. This method enables one to evaluate the rate constant for a first order reaction when the initial and final (infinity) concentrations of reactants are not known.

Procedure: Prepare 0.1 N NaOH solution. Pipette out 50 cm^3 in a stoppered bottle and thermostat it for 15 minutes to attain equilibrium temperature. Add 1 cm^3 of diacetone alcohol, mix well and then transfer into the dilatometer. Fill the bulb of the dilatometer without any air bubble. Make sure that the level of the solution is at the lowest end of the capillary tube. Record the readings of the meniscus every 2 min for a period of 60 minutes and tabulate the results.

Treatment of Results: Tabulate the dilatometer readings for the decomposition of diacetone alcohol as shown below.

Time (min)	Dilatometer Reading (v)	Time (min)	Dilatometer Reading (v′)	(v – v′) Δt = 30 min	ln (v – v′)
2	32				
4	34				
6	36				
8	38				
10	40				
12	42				
14	44				
16	46				
18	48				
20	50				
22	52				
24	54				
26	56				
28	58				
30	60				

The observed readings may be segmented into two sets, readings (v) in the first 30 minutes and those (v′) from 30 – 60 minutes. The difference (v – v′) between the corresponding readings from the two sets is noted. Then plot log (v – v′) versus t; from the slope of the line, calculate the rate constant.

Result:

Report the value of the rate constant obtained.

Note:

Repeat the experiment with another quantity of diacetone alcohol and comment on the value of the rate constant obtained.

EXPERIMENT 41
Rate of Decomposition of Ammonium Nitrite (or Benzene diazonium chloride)

Aim: To study the rate of decomposition of ammonium nitrite or benzene diazonium chloride in aqueous solution.

Apparatus required: 100 cm^3 (or high capacity) burette, gas collection tubes, trough, stand.

Chemicals Required: Ammonium nitrite or benzene diazonium chloride.

Principle: The decomposition reaction of NH_4NO_2 occurs at around 70°C while that of benzene diazonium chloride may be followed at room temperature itself.

$$NH_4NO_2 \xrightarrow{60-70°C} 2H_2O + N_2$$

$$C_6H_5N_2^+Cl^- + H_2O \longrightarrow C_6H_5OH + HCl + N_2$$

Both these reactions follow first order kinetics and are attended with the evolution of nitrogen gas. The liberated nitrogen gas can be collected over water using an inverted burette filled with water in a trough containing water. From the amount of the gas collected at different time interval, the rate of decomposition of the material can be calculated.

Procedure: Set up the reaction apparatus as shown in the (Fig. 23) and place a known amount of ammonium nitrite (it requires heating) or benzene diazonium chloride. Note down the volume of nitrogen evolved as a function of time. Record the laboratory temperature and pressure as well as the height of water column at every time of reading. Allow the reaction to go to completion and record the total volume of nitrogen evolved. The total volume of nitrogen evolved at the completion of reaction reduced to NTP conditions (this need not be necessarily done since the active pressure at which nitrogen is collected is close to that of the atmospheric pressure, since even a 100 cm of water column will affect the pressure by one or two cm Hg column) will be proportional to the initial concentration of ammonium nitrite or the benzene diazonium chloride taken.

The difference between this volume at the completion of the reaction, say V_∞, and the volume at any time t, V_t is proportional to the amount of ammonium nitrite or benzene diazonium chloride remaining. The first order rate constant can be calculated by substituting the recorded volume quantities in the expression $k = (2.303/t) \times \log \{V_\infty/(V_\infty - V_t)\}$.

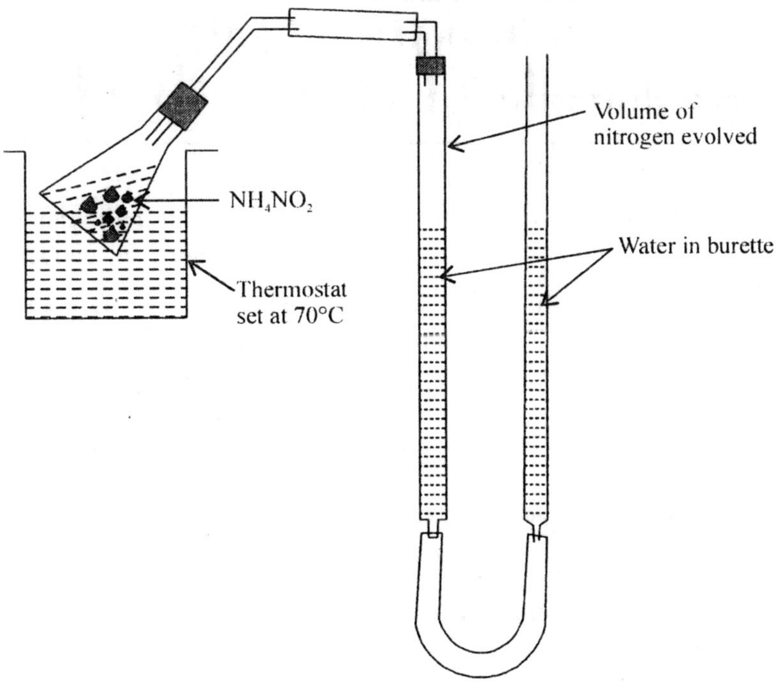

Fig. 23. Typical experimental arrangement for studying decomposition of NH_4NO_2 (or $C_6H_5N_2^+Cl$).

Treatment of Data: Record the volume readings as a function of time as shown in the table.

Temperature: °C

Time (min)	Volume of nitrogen evolved (cm^3)	k min^{-1}
0	V_0 =	
2		
4		
6		
8		

Time (min)	Volume of nitrogen evolved (cm^3)	k min^{-1}
10		
15		
20		
30		
40		
∞	V_∞	

Mean k = ...

Draw a graph of $\log(V_\infty - V_t)$ against time. From the slope of the resulting line, the rate constant value can be evaluated.

Result:

Report the value of the first order rate constant for the decomposition of ammonium nitrite or benzene diazonium chloride.

Note:

However, it is not always convenient to record the volume of nitrogen evolved at the completion of reaction. The reasons are (i) the volume of nitrogen evolved at the completion of the reaction may be higher than the gas burette that one uses to collect nitrogen and (ii) the time taken for the completion of the reaction is rather long even at higher temperatures. Under these circumstances, one can still calculate the value of the first order rate constant using Guggenheim method.

EXPERIMENT 42
Evaluation of Rate Constant by Guggenheim Method

Aim: To evaluate the rate constant for the hydrolysis of an ester by the Guggenheim method.

Apparatus Required: 250 cm^3 conical flasks, burette, pipettes and reaction bottles.

Chemicals Required: Distilled ester (methyacetate), 0.5 M H_2SO_4, 0.2 M NaOH, phenolphthalein indicator and ice or ice water.

Principle: For a first order reaction the rate constant is given by the equation,

$$k_1 = (2.303/t) \log \{a/(a - x)\}$$

where a and (a – x) are the initial concentration and that at time t of the ester or in the exponential form it can be given as

$$(a - x) = ae^{-kt}$$

When the initial and the final concentrations are not known as in the case of a radioactive decay process or in a very slow reaction, the Guggenheim method is adopted. According to this, the reaction is to be followed at convenient but fixed time intervals, say t. Then the reaction is repeated at another fixed time intervals, say t + Δ t. (For better results, it is to be chosen such that the time interval is two or three times the half life period of the reaction). If the initial concentration is a and x_1 and x_2 are the amounts reacted in time t and t + Δ t, from the first order rate equation we can write,

$$a - x_1 = a_0e^{-kt} \quad \text{for readings at time t}$$

and

$$a - x_2 = a_0e^{-k(t + \Delta t)} \quad \text{for readings at time } t + \Delta t$$

Subtracting one from the other and rearranging in the logarithmic form we can see that $\log (x_1 - x_2)$ versus time plot will be linear, whose slope can be used to calculate the rate constant.

Procedure: Take 95 cm^3 of 0.5 M H_2SO_4 in a reaction bottle and add 5 cm^3 of the distilled methylacetate to it. Shake well and pipette out 5 cm^3 of this reaction mixture into a conical flask containing a few grams of ice or ice water. Add 2 drops of phenolphthalein indicator. Titrate this solution with the 0.2 M NaOH solution taken in the burette till a pale permanent pink colour is seen. Note the volume of the alkali consumed for the titration. Repeat this procedure exactly at 15 minutes intervals and collect at least 12 readings, by following the reaction for three hours.

Treatment of Data: Tabulate the readings as shown:

S. no.	Time	Volume NaOH (cm^3)	$x_1 - x_2$	$\log (x_1 - x_2)$
1	15			
2	30			
3	45			
4	60			
5	75			
6	90			
7	105			
8	120			
9	135			
10	150			
11	165			
12	180			

The data in the table are divided into two sets – from S. No. 1 to 6 as set I and from S.No. 7 to 12 as set II, so that the time interval between any two readings is 90 minutes.Find out the $x_1 - x_2$ values by subtracting the volume of NaOH from set II from the corresponding values in set I. Plot the $\log (x_1 - x_2)$ values thus calculated against the time and from the slope of the resulting straight line calculate the rate constant.

Result:

The rate constant for the hydrolysis of methylacetate = ... min^{-1}.

EXPERIMENT 43
Effect of Ionic Strength on the Rate of Persulphate-Iodide Reaction

Aim: To study the effect of ionic strength on the rate of oxidation of iodide ion by potassium peroxydisulphate.

Apparatus Required: Burette, pipettes, reaction bottles and conical flasks.

Chemicals Required: Potassium iodide crystals, potassium nitrate, potassium peroxydisulphate ($K_2S_2O_8$), sodium thiosulphate, starch indicator and ice or ice cold water.

Principle: The net reaction under consideration is

$$S_2O_8^{2-} + 2\,I^- \longrightarrow 2\,SO_4^{2-} + I_2$$

As it involves ions, this reaction is highly susceptible to any variation in the ionic strength of the reaction medium. The kinetics of this reaction is followed by determining the concentration of the liberated iodine volumetrically by reacting the aliquot of the reaction mixture at different time intervals with a standard sodium thiosulphate solution to a starch end point. By repeating the reaction at different concentrations of the initially added neutral salt like KNO_3 and evaluating the rate constant, we can understand the effect of the ionic strength on the rate of this reaction using the Bjerrum – Bronsted equation. This equation is given as

$$\log k = \log k_o + 1.02\, Z_A Z_B \sqrt{\mu}$$

where k and k_o are the rate constants in the presence and absence of the added salt respectively, Z_A and Z_B are the charges on the two reactants respectively and the μ is the ionic strength.

Procedure: Weigh 0.1 g of solid KNO_3 and place it in a clean reaction bottle. Pipette out 50 cm^3 of 0.05 M $K_2S_2O_8$ solution into the reaction bottle and place it in a trough of water. Pipette out 50 cm^3 of 0.1 M KI solution into the reaction bottle, noting the time (or starting a stop watch) when the pipette is half run. Stopper the reaction bottle, shake the contents well and keep it back in the trough. Pipette out 10 cm^3 aliquot of the reaction mixture immediately (noting the time when the pipette is half-drained) into a conical flask containing some ice or about 5 cm^3 of ice-cold water. Titrate against 0.01 M sodium thiosulphate solution using starch as the indicator. Repeat withdrawing the aliquot at 10 minutes interval initially and at 15 minutes interval later, titrate and find the $[I_2]$ liberated at different intervals of time.

Redo the whole of the above procedure by taking 0.2 g and again with 0.3 g of KNO_3 in the reaction bottle (instead of 0.1 g).

Treatment of Data: Tabulate the observed data in the following format and calculate the rate constant.

$[K_2S_2O_8]$ = 0.025 M(a); [KI] = 0.05 M(b); $[S_2O_3^{2-}]$ = 0.01 M;

KNO_3 = 0.1 g.; Temperature = °C.

S. no.	Time (s)	Titre (cm^3)	$[I_2]$ = (V × 0.01) M(x)	(a – x) M	k_2
1					
2					
3					
4					
5					
6					
7					

Mean k_2 = ... $dm^3\ mol^{-1}s^{-1}$.

Draw similar tables for the data collected with 0.2 g and 0.3 g of KNO_3.

In each case the second order rate constant k_2 is calculated using the equation

$$k_2 = (1/2at)\ \{x/(a - x)\}$$

where t is the time.

The value of k_2 can also be calculated from the slope of the straight line obtained by plotting the x/(a – x) values on the y-axis and the time on the x-axis.

Calculate the concentration of KNO_3 added in each experiment by dividing the mass of the salt added by its molecular weight. As it is a uni-univalent electrolyte, its concentration(c) is equal to its ionic strength. Hence plot the log k_2 values on the y-axis and the $\sqrt{\mu}$ on the x-axis. As expected from the Bjerrum–Bronsted equation, a straight line (Fig. 24) will result. From the slope of the line, the nature of the reacting species can be inferred.

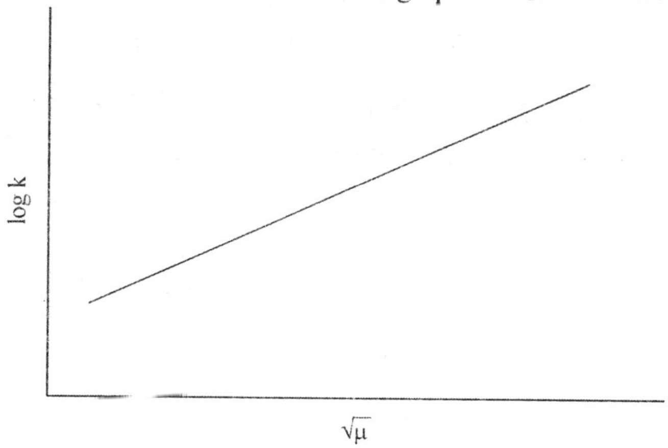

Fig. 24. Bjerrum-Bronsted plot.

Results:

(i) Report the k_2 values at the different concentrations of the neutral salt KNO_3 added.

(ii) Comment on the nature of the reacting species

Note:

Studies on such salt effect on the rate of the reaction are useful in deciding the nature of the reactive species. From the Bjerrum – Bronsted equation it can be deduced that when the logarithm of the rate constant values are plotted against the square root of the ionic strength, the positive slope for the resulting straight line indicates that in a binary reaction both the reacting species have the same charge; a negative slope for the line indicates the reacting species carry opposite charges and a straight line parallel to the ionic strength axis (slope = 0) suggests that at least one of the reacting species is neutral.

EXPERIMENT 44

Kinetics of Oxidation of Alcohol by Potassium Dichromate – Spectrophotometry

Aim: To determine the specific rate constant for the oxidation of ethanol by potassium dichromate.

Apparatus Required: Spectrophotometer (or a colourimeter); volumetric flasks 100 or 50 cm^3, pipettes 10 cm^3 and 2 cm^3 – graduated.

Chemicals Required: Potassium dichromate crystals, pure ethanol, 2 M H_2SO_4.

Principle: The reaction may be represented as

$$3\ C_2H_5OH + Cr_2O_7^{2-} + 8\ H^+ \longrightarrow 3\ CH_3CHO + 2\ Cr^{3+} + 7\ H_2O$$

The reaction is first order in each of the reactants. As the reactant, $Cr_2O_7^{2-}$ is coloured and it disappears in the course of the reaction, its concentration may be followed with time using a spectrophotometer (or a colourimeter) at 420 nm, the wave length at which $Cr_2O_7^{2-}$ obeys Beer– Lambert's law.

Procedure: Prepare 0.5 M ethanol solution and 0.1 M potassium dichromate solution by dissolving appropriate quantities in distilled water. Pipette out 0.5 cm^3 of 1 M ethanol solution into a cell containing 1 cm^3 of 2 M H_2SO_4. Add 0.2 cm^3 of 0.1 M $K_2Cr_2O_7$ solution and 0.3 cm^3 distilled water to this mixture and quickly shake the contents. Place the cell in the cell compartment of the spectrophotometer (or the colourimeter). Follow the absorbance of the reaction mixture at 420 nm, keeping water as reference, at 1 minute interval, till the absorbance reduces to at least one-fourth of the initial value.

Treatment of Data: Tabulate the values as shown in the table.

Time (min)	1	2	3	4	5	6	8	10	12	15	18	21	24	27	30
Absorbance (a. u.)															

From the data in the above table, the first order rate constant may be calculated as

$$k_1 = (2.303/t) \log \{A_0/A_t\}$$

when A_o and A_t are the absorbances measured at the commencement of the reaction and at various time intervals respectively.

Also, plot the absorbance at different time intervals against time; a straight line is obtained. From the slope of this line, the specific rate constant (k_1) can be calculated as, $k_1 = -2.303 \times \text{slope}$.

Result:

Report the value of the specific rate constant in the units of min^{-1} or s^{-1}.

Note:

The solutions may be proportionally diluted as per the maximum absorbance limit of the available instrument.

PART VI

Surface Chemistry

EXPERIMENT 45

Adsorption Characteristics of Acetic Acid on Charcoal

Aim: To study the adsorption of acetic acid by activated charcoal from an aqueous solution.

Apparatus Required: Conical flasks, funnel, well cleaned and dry stoppered bottle, 5 cm^3 pipette, 50 cm^3 burette, glass rod, standard flask (250 cm^3).

Chemicals Required: 0.5 N, 0.4 N, 0.3 N and 0.2 N acetic acid, standard sodium hydroxide solution, phenolphthalein indicator, activated charcoal.

Principle: Adsorption is a surface phenomenon. When charcoal powder is added to a dilute solution of acetic acid, it is adsorbed on the surfaces of charcoal particles. Generally, the adsorption increases with increase in the surface area of the adsorbent. The amount of adsorbate adsorbed over unit mass of adsorbent (A) is related to the concentration (C) of adsorbate by the Freundlich isotherm as, $A = kC^n$, where k and n are constants.

Procedure: Pipette out 30 cm^3 of glacial acetic acid and make up to 500 cm^3 so that the concentration is approximately 1 N. Standardize this solution using standard NaOH (0.4 N) solution using phenolphthalein as indicator. Dilute this standardised acetic acid appropriately and prepare solutions of concentrations 0.5, 0.4, 0.3, and 0.2 N. Weigh out accurately 1.0 g of activated charcoal and transfer into a well cleaned dry 250 cm^3 stoppered bottle. Then add 200 cm^3 of 0.5 N acetic acid into the bottle containing activated charcoal. Shake the mixture intermittently and place at room temperature (about 2 hours). After equilibrium is reached filter the solution. Pipette out 10 cm^3 of the filtrate and titrate against standard NaOH solution using phenolphthalein as indicator. Repeat the above procedure for 0.4 N, 0.3 N and 0.2 N solutions of acetic acid.

Treatment of Results: Record your observations in a table as shown below.

Weight of charcoal, (m) = 1.0 g; [NaOH] = (N_1); Temperature = °C

Initial [HOAc] (C_0)	Titre value ml, (V)	$[HOAc]_{free} = V \times N_1/10$ (C)	$[HOAc]_{absorbed}$
0.2			
0.3			
0.4			
0.5			

1. Calculate the mass of acetic acid adsorbed by charcoal (x grams).
2. Plot according to the Freundlich isotherm

$$\log (x/m) = \log k + n \log c$$

where c = $[HOAc]_{adsorbed}$.

3. Knowing the weight of the adsorbent (m), draw a graph taking log(x/m) on the y-axis against log $[HOAc]_{adsorbed}$ on the x-axis. Calculate the constants n and k respectively from the slope and intercept of the line.

Result:

The Freundlich isotherm is verified as a straight line is obtained in the log (x/m) versus log [HOAc] plot.

The value of the constants, n = . . . and k = ...

EXPERIMENT 46

Surface Tension of a Liquid Using Stalagmometer

Aim: To determine the surface tension of a given liquid by drop number method using a Stalagmometer.

Apparatus Required: A wide mouthed bottle fitted with a rubber stopper with two holes through which a stalagmometer and tube can be passed, clamp stand, thermometer, thermostat, distilled water and the liquid whose surface tension has to be determined.

Principle: The method is based on the principle that when a liquid is allowed to pass through a capillary tube, held vertically, such that the drop falls under its own weight (and not forced down), then the weight of the drop is proportional to the surface tension of the liquid. If two liquids with surface tensions γ_1 and γ_2 are passed through the same capillary then $\gamma_1/\gamma_2 = w_1/w_2$, where w_1 and w_2 are the mean weights of the drops falling off from the capillary.

Procedure: Clean the stalagmometer and the receiving bottle with necessary solvents and if necessary, with chromic acid; wash with distilled water. Connect a rubber tubing at the top of the stalagmometer (Fig. 25) with a screw pinch cock. Immerse the lower end of the stalagmometer tube in distilled water and suck through the rubber tubing until the water level rises above or upto the mark x. Close the screw pinch cock and fit the rubber stopper carrying the stalagmometer and glass tube to the receiving bottle. Place the assembly in a thermostat. After attainment of the thermal equilibrium, slowly open the screw pinch cock and allow the drops of distilled water to fall down. Count the number of drops as the liquid meniscus falls from the mark x to y in the stalagmometer. Now repeat the experiment with the given liquid after cleaning and rinsing the stalagmometer. Also determine the density of distilled water and that of the liquid using a specific gravity bottle at the experimental temperature.

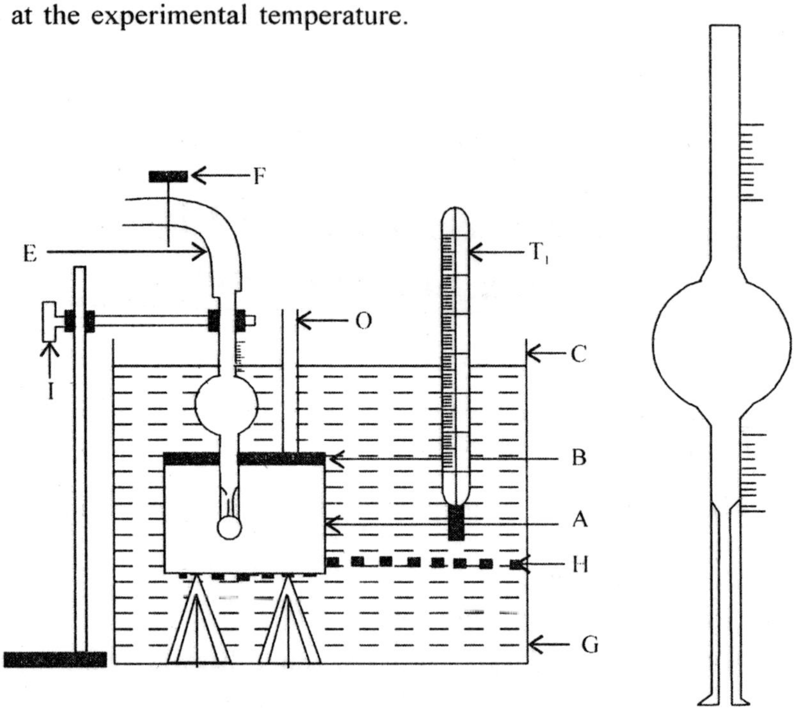

Fig. 25. Measurement of surface tension.

A — Wide mouth receiving bottle
B — Tightly fitted rubber stopper
C — Stalagmometer
D — Glass tube
E — Rubber tubing
F — Screw-Pinch-cock
G — Thermostat
H — Liquid under test
I — Clamp
T_1 — Thermometer

Treatment of Data:

Surface tension of water at T°C $= \gamma_2$ (note down from standard tables).

Weight of empty specific gravity bottle $= w_1$

Weight of empty specific gravity bottle + water $= w_2$

Weight of empty specific gravity bottle + liquid $= w_3$

Density of liquid/density of water $= d_1/d_2 = (w_3 - w_1)/(w_2 - w_1)$

Surface tension of the given liquid $= r_2\,(d_1/d_2)\,(n_1/n_2)$

$= r_2(w_3 - w_1)/(w_2 - w_1)\,(n_2/n_1)$

Result:

Report the surface tension of the liquid in the unit of dyne/cm or Newton/metre.

EXPERIMENT 47

Critical Micelle Concentration from Surface Tension Measurements

Aim: To determine the critical micelle concentration of a surface active material by surface tension method.

Apparatus Required: Stalagmometer (or Du Nouy's tensiometer), 100 cm^3 volumetric flasks, pipettes.

Chemicals Required: Sodium laurylsulphate (SLS), distilled water.

Principle: Substances that get adsorbed on to a solid or liquid surface and decrease the surface tension of the medium are known as surface active materials or surfactants. Soaps and detergents are good examples for surfactants. The aqueous solutions of such substances are colloidal in nature and with increase in concentration, they tend to aggregate. These aggregates are know as association colloids or micelles.

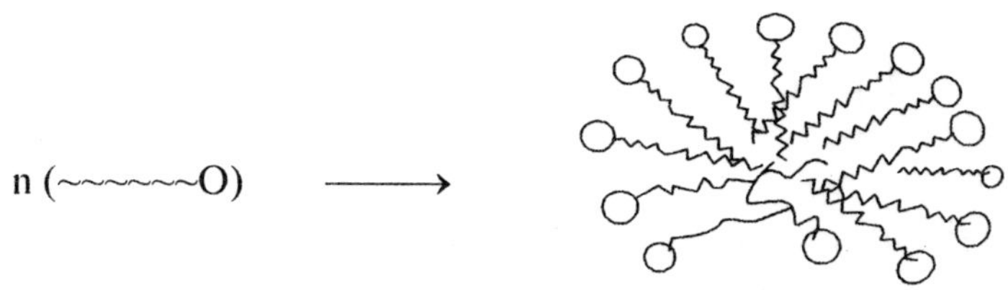

surfactant molecule micelles

The concentration above which these micelles become detectable is called the critical micelle concentration (cmc). Consequently, with increase in concentration of the surfactant the surface tension of water progressively decreases and becomes constant above the cmc.

Procedure: Weigh 0.576 g of sodium laurylsulphate [its molecular formula is $CH_3 - (CH_2)_{10} - CH_2{-}\ SO_4^-\ Na^+$.], dissolve it in water and make up to 100 cm^3 in a volumetric flask. This solution corresponds to 0.02 M. Dilute this solution appropriately and prepare the following concentrations. 0.01 M, 0.008 M, 0.006 M, 0.005 M, 0.004 M, 0.002 M, 0.001 M. Determine the surface tension of each of these solutions using the stalagmometer, as described in *experiment 46*. Tabulate the results as shown in the table.

[SLS] M	0.001	0.002	0.004	0.005	0.006	0.008	0.01	0.02
S. T. N/m								

Plot the values of surface tension against concentration of [SLS]. The resulting graph will be as shown in (Fig. 26). Draw the two straight lines and extend them to find the point of intersection. The concentration corresponding to this point gives the critical micelle concentration of the surfactant.

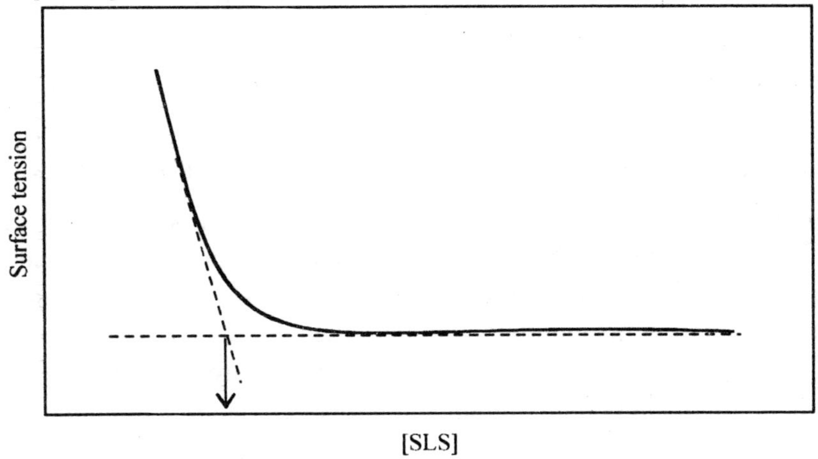

Fig. 26. Critical micelle concentration of SLS.

Result:

Report the critical micelle concentration (cmc) of sodium laurylsulphate as determined by surface tension method.

PART VII
Photometry

EXPERIMENT 48

Verification of Beer and Lambert's Law

Aim: To verify the Beer-Lambert's Law and to determine the unknown concentration of a given solution.

Apparatus Required: Spectrophotometer (or photocolourimeter), cells, standard flasks, beakers and pipettes.

Chemicals Required: 0.01 M $KMnO_4$ solution in distilled water (or $K_2Cr_2O_7$ solution), distilled water.

Principle: The Beer-Lambert's law states that when a beam of monochromatic light passes through a solution (containing a species that absorbs in that wavelength region), the decrease in the intensity of the radiation transmitted is directly proportional to the concentration of the solution taken and the path length of the solution.

The logarithm of the ratio of the intensity of the incident light to that of the transmitted light, called the absorbance (A) is proportional to the concentration (c) of the solution and the path length (l). i.e.,

$$\log (I_0/I) \propto c\, l$$

or

$$A = \varepsilon\, c\, l$$

where ε is referred to as the molar extinction coefficient which is characteristic of the chemical species absorbing the radiation.

Therefore a plot of absorbance versus concentration will be a straight line from which one can determine the value of the molar extinction coefficient of the chemical species absorbing the radiation.

This law is valid in dilute solutions since deviations occur to this linear variation at higher concentrations due to ion association and other phenomena occurring in concentrated solutions.

Procedure: Prepare 0.1 N solution of $KMnO_4$ by dissolving 0.316 g of crystals of $KMnO_4$ in 100 cm^3 of distilled water. Dilute this stock solution with distilled water and get 0.1, 0.2, 0.4, 0.8, 1 and 2 milli-molar solutions of $KMnO_4$.

Use one of these solutions and record the spectrum or determine the absorbance for this solution at wavelengths from 400 to 600 nm. Note the wavelength at which maximum absorbance is observed. This is the λ_{max}. Now set the instrument to this wavelength and measure the absorbance of each of the solutions at this wavelength.

Treatment of Data:

1. Determination of λ_{max}

Wavelength (nm)																			
Absorbance (a.u.)																			

2. Absorbance for different concentrations of $KMnO_4$ solutions at the λ_{max}.

$[KMnO_4]$ mM	0.1	0.2	0.4	0.6	0.8	1.0	2.0
Absorbance (a.u.)							

3. Determination of molar extinction coefficient: Plot a graph of the absorbance versus concentration. The graph is a straight line passing through the origin. Find the slope of the straight line. Slope is equal to εt and since generally, the width of the cell employed is known 1 cm, slope = ε. Compare this value with the value reported in literature.
4. Verification of the validity of Beer – Lambert's Law: The plot of absorbance versus concentration will be a straight line upto a particular value of concentration. This linear portion indicates the range of the validity of the Beer Lamberts law. The point at which deviation, if any, occurs from linearity indicates the concentration range above which this law does not hold good.
5. Determination of the concentration of the test (unknown) solution: At the same value of λ_{max}, determine the absorbance of the given unknown concentration solution. By interpolating this absorbance value for the test solution in the calibration line the concentration of the test solution can be read.

Results:

1. Report on the value of the molar extinction coefficient. (Note the magnitude of this value and comment on the value)
2. State the limits of the values of the concentration of the solution up to which the Beer Lambert's law obeys.
3. State the value of the concentration of the unknown solution. Comment on the error of the value obtained.

Note:

Any coloured solution can be used to determine the validity of the Beer-Lambert's law. However, the correct λ_{max} has to be selected each time since λ_{max} is a function of the nature of the chemical species involved. For example solution of copper ammine complex can be used for verifying Beer-Lambert's law.

EXPERIMENT 49

Job's Continuous Variation Method

Aim: To find the stoichiometry of a complex formed between a metal ion and a given ligand by Job's continuous variation method and also to determine the stability constant of the complex formed.

Apparatus Required: spectrophotometer or colourimeter with suitable cells, standard flasks, pipettes and test tubes.

Chemicals Required: 0.001 M and 0.002 M of HCl, 0.001 M salicylic acid and 0.001 M ferric nitrate solution.

Principle: Job's Method, also called the Method of Continuous Variation, is a simple and effective approach for the determination of the stoichiometry of a chemical reaction or the stoichiometry of a complex formed. Let us consider a general reaction

$$a\,A + b\,B \rightleftharpoons d\,D \quad (1)$$

which can be rewritten in the following form by dividing all coefficients by "a" as

$$A + k\,B \rightleftharpoons m\,D \quad (2)$$

where k = b/a and m = d/a.

Job's method is based on the principle that if a series of solutions is prepared, each containing equal number of moles of A and B, but a different ratio, R, of moles B to moles of A, the maximum amount of product, D is obtained in the solution at which R = k (the stoichiometric ratio).

To implement Job's Method experimentally, one prepares a series of solutions containing a fixed total number of moles of A and B, but in which the R is systematically varied from large to small, and measures the amount of product obtained in each solution. One then plots amount of product versus R, and obtains a maximum at the initially-unknown value of k.

That the maximum amount of product should occur at the stoichiometric ratio can be justified mathematically. (See the note below.)

Procedure: Prepare 0.001 M solution of ferric nitrate (A) and also the same concentration of salicylic acid (B).

Prepare various mixtures by mixing different volumes of the two solutions but keeping the total volume of the mixture constant, say 1 mL of A + 9 mL of B, 2 mL of A + 8 mL of B, etc.

Using appropriate solution (say in this case even a 5 + 5 cm^3 solution is sufficient) determine the wavelength of maximum absorption λ_{max}. At this wavelength determine the absorbance of each of the solutions given above. Plot the absorption as a function of the composition. The ratio of the reagents corresponding to the maximum absorption in the curve gives the composition of the complex.

Tabulation of Data:

Fe^{3+} solution (cm^3)	1	2	3	4	5	6	7	8	9
Salicylic acid	9	8	7	6	5	4	3	2	1
Absorbance									

Alternatively, a *slope ratio method* can also be adopted. In this case, two sets of solutions are prepared as follows:

In the first set of five solutions, keep the concentration of the metal ion constant and vary the concentration of the ligand; in the second set keeping the concentration of the ligand constant vary the concentration of the metal ion solution but make sure all the solutions have the same final volume by using HCl of appropriate (about 0.002 N) concentration.

Determine the absorbance of each of these solutions. For these two sets of data plot the absorbance versus the volume of metal ions or the volume of the salicylic acid added. From the ratio of the two slopes one can determine the stoichiometry of the complex formed.

The stability constant determination requires the estimation of the concentration of the complex and also the concentration of the free metal ions and free ligand. In the example under consideration using ferric ions and salicylic acid, consider one has determined the absorption at the λ_{max} of the complex. Also assume the value of the molecular extinction coefficient of the complex, ε, has already been determined. So if one were to know the concentrations of the metal ion and ligand taken and also if the concentration of the complex were to be determined from the value of absorbance using the formula $A = \varepsilon\ l\ [\text{complex}]$, the free metal ion concentration and free ligand concentration can be calculated if the stoichiometry of the complex is known.

The stability constant of the complex can then be calculated

$$K = [\text{complex}]/[Fe^{3+}]^{a}_{\text{free}}\ [\text{ligand}]^{b}_{\text{free}}$$

Since in our example it is a 1:1 complex

$$[Fe^{3+}]_{\text{free}} = [\text{salicylic acid}]_{\text{free}} = [\text{total iron taken}] - [\text{complex}]$$

$$K = (A/\varepsilon l)/\{[Fe^{3+}]_{\text{total}} - (A/\varepsilon l)\}^{2}$$

Result:

Report (i) the value of ε of the complex formed (ii) the stoichiometry of the complex and (iii) the value of K. (Comment on your results indicating the probable error in the values of ε and K).

Note 1:

The mathematical justification is also quite simple. We use variable "x" to represent the moles of A in a particular solution, and assume that the total moles of A and B is to be kept at 1.0 throughout the series of solutions. Then in each solution it will be true that

$$\text{moles A} = x \qquad \text{and} \qquad \text{moles B} = 1 - x$$

Our goal is to show that the maximum amount of product is obtained when R = moles B/moles A = $(1 - x)/x$ is equal to k. We approach this by finding the value of x that maximizes product.

According to equation (2), if x is less than the stoichiometrically correct amount of A, then A is limiting and moles product = mx. A plot of moles product versus x over a series of solutions should be linear, with slope m. Similarly, if x exceeds the stoichiometrically correct amount of A, then B is limiting and moles product = m $(1 - x)/k$. A plot of moles product versus x over a series of solutions should also be linear, with slope = – m/k. The first plot will proceed up to the right as x increases. The second plot will proceed down to the right as x increases. At some point then, the two straight lines will intersect. At the intersection, they have a point in common. The value of x corresponding to this point is obtained by equating the ordinate values and solving for x:

$$mx = m(1 - x)/k$$

Solving gives $x = 1/(1+k)$.

Substituting in the expression for R, $(1 - x)/x$, we find that the two lines intersect when

$$R = (1-x)/x = \{1 - [1/(1+k)]\}/[1/(1+k)] = k$$

Because the amount of product increases as k is approached from either direction, the point of intersection of the lines occurs at the maximum amount of product obtainable. It is clear that maximum product is obtained when $R = k$.

***Note 2* :**

The above experiment may also be conducted using the following sets of solutions.

(i) Dilute HCl, dilute sulphosalicylic acid in HCl, dilute ferric ammonium sulphate solution in HCl; the concentration of each of these solutions must be known; and

(ii) Dilute solution of nickel nitrate and also dilute 1, 10-phenanthroline solution; the concentration of these two solutions should be known. (the stoichiometry obtained is 1: 3 in this case).

EXPERIMENT 50

Simultaneous Estimation of Manganese and Chromium in a Solution of Dichromate and Permanganate Mixture

Aim: Determine simultaneously the concentration of dichromate and permanganate ions in a given mixture by colorimetric method.

Apparatus Required: Colourimeter or spectrophotometer with cells, standard flasks, burettes, pipettes.

Chemicals Required: 0.01 N $K_2Cr_2O_7$ solution and 0.01 N $KMnO_4$ solution prepared in 1 N sulphuric acid as stock solution.

Principle: The individual concentration of two species present in a mixture can be determined by colorimetry only if both the species in the solution obey Beer's law over their concentration range present in the mixture and if the species do not interact among themselves. The absorbances are additive at a given wave length

i.e.,

$$A_{\lambda_1} = A_1 + A_2 \text{ (at wave length } \lambda_1)$$

$$A_{\lambda_2} = A_1 + A_2 \text{ (at wave length } \lambda_2)$$

where A_{λ_1}, A_{λ_2} are the absorbances of the mixture at the respective wave lengths; A_1 and A_2 are the absorbance of the individual species 1 and 2 at the respective wave length.

In this measurement λ_1 and λ_2 are shown such that substance 1 absorbs strongly in one wavelength and the substance 2 absorbs strongly in the other wavelength.

According to Beer Lambert's Law

Absorbance = εcl and if one were to use a 1 cm cell then, absorbance = εc, where ε is the molar absorptivity and c is the concentration of the absorbing species.

$$A\ (\text{at}\ \lambda_1) = \varepsilon_1\ c_1\ (\text{at}\ \lambda_1) + \varepsilon_2\ c_2\ (\text{at}\ \lambda_1);$$

$$A\ (\text{at}\ \lambda_2) = \varepsilon_1\ c_1\ (\text{at}\ \lambda_2) + \varepsilon_2\ c_2\ (\text{at}\ \lambda_2);$$

If one were to estimate the values of all the four molar extinction coefficients namely ε_1 at λ_1 and λ_2 and thc value of ε_2 at λ_1 and λ_2 then one can write the solutions for c_1 and c_2 as follows

$$c_1 = [\ \varepsilon_2(\text{at}\ \lambda_2)\ A\ (\text{at}\ \lambda_1) - \varepsilon_2(\text{at}\ \lambda_1)\ A\ (\text{at}\ \lambda_2)]/[\ \varepsilon_1\ (\text{at}\ \lambda_1)\ \varepsilon_2\ (\text{at}\ \lambda_2) - \varepsilon_2\ (\text{at}\ \lambda_1)\ \varepsilon_1\ (\text{at}\ \lambda_2)]$$

$$c_2 = [\varepsilon_1\ (\text{at}\ \lambda_1)\ A\ (\text{at}\ \lambda_2) - \varepsilon_1\ (\text{at}\ \lambda_2)\ A\ (\text{at}\ \lambda_1)]/[\varepsilon_1\ (\text{at}\ \lambda_1)\ \varepsilon_2\ (\text{at}\ \lambda_2) - \varepsilon_2\ (\text{at}\ \lambda_1)\ \varepsilon_1\ (\text{at}\ \lambda_2)]$$

Procedure: Prepare 0.008, 0.006, 0.004, 0.002 and 0.001 N solutions of $K_2Cr_2O_7$ and $KMnO_4$ in 1 N sulphuric acid as solvent, using appropriate stock solutions. Select λ_{max} for both dichromate and permanganate using the 0.008 N solutions. The λ_{max} for dichromate will be found to be 440 nm – this is λ_1; λ_{max} for permanganate will be 525 nm – this will be λ_2.

Record the absorbance for all the solutions prepared, both for dichromate and permanganate at 525 nm and also at 440 nm. Also record the absorbance of the test solution containing a mixture of dichromate and permanganate at both the wavelengths.

Treatment of Data: If the absorbance values are plotted against concentration one will get two straight lines for each of the wave lengths used. From the slope of these lines one can calculate the values of molar extinction coefficients. From this one can calculate the concentration of the two species in the unknown solution.

One can prepare a series of mixed solutions 10 ml of dichromate and 40 ml of permanganate [20 + 30], [30 + 20] and [40 + 10] and tabulate the data as follows

Volume of $K_2Cr_2O_7$ solution (cm^3)	Volume of $KMnO_4$ solution (cm^3)	Absorbance observed	Absorbance calculated
50	—		
40	10		
30	20		
20	30		
10	40		
—	50		

Since concentrations are known and molar extinction coefficients are also known, one can calculate the value of optical density and compare the calculated values with that observed at each of the wave lengths.

This provides a means to establish the operation of the additivity principle.

Result:

Report the concentration of each of the components and also comment on the accuracy of the result reported.

EXPERIMENT 51

Spectrophotometric Determination of Critical Micelle Concentration

Aim: To determine the critical Micelle concentration of a surfactant.

Chemicals Required: Sodium laurylsulphate (or any surfactant), benzoylacetone, distilled water.

Apparatus Required: Spectrophotometer with UV region, volumetric flasks and pipettes.

Principle: Surfactants are amphiphilic molecules capable of forming aggregates called micelles above a certain minimum concentration called Critical Micelle Concentration (CMC). The anionic surfactant, sodium laurylsulphate (SLS) forms micelles with hydrocarbon core and ionic surface. Consider a diketone like Benzoylacetone (BZA). The diketones exist in two tautomeric forms, namely the ketonic and enolic forms as shown below.

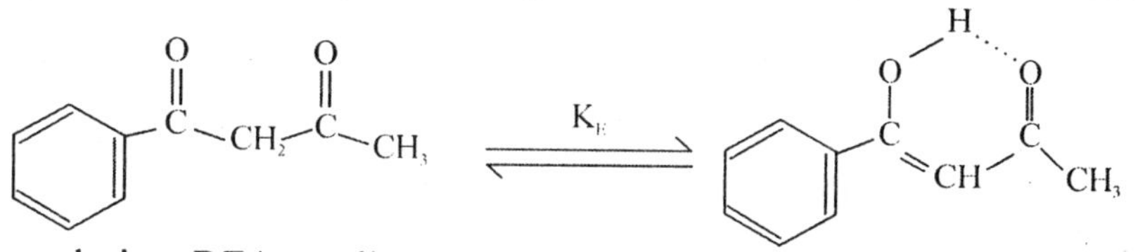

In solution BZA enolizes practically exclusively to cis enolic form which exists in a cyclic configuration stabilized by intramolecular hydrogen bonding. Thus, the keto-enol equilibria of BZA is extremely solvent sensitive and the proportion of the enolic form is greater in nonpolar solvents such as cyclohexane than in polar solvents like water. BZA exists in water as a mixture of both tautomeric forms (enol form is around 37.5%). When surfactant is added to a water solution of BZA the amount of enol increases abruptly when the surfactant concentration rises above CMC because the enolic form is taken up by the micelles which provide a less polar environment compared to the solvent. If one were to measure the absorption spectrum as a function of concentration and monitors the absorbance at the wavelengths 312 nm (absorption maximum for the enolic form) and at 250nm (absorption maximum for the keto form) then there will be sudden gradient change around the CMC.

Procedure: Prepare a standard solution (~0.03 mol L^{-1}) of BZA in dioxane or any other water miscible nonpolar solvent. From this stock solution prepare an aqueous BZA solution by pipetting 0.40 mL into a 25 mL standard volumetric flask and make up the volume. For each measurement pipette 0.40 mL of the aqueous solution of BZA to a 1.0 cm quartz cell together with the appropriate amount of surfactant (additive if any) and water so that the final volume of the mixture is 3 mL.This means the approximate concentration of BZA in this solution used for measurement will be around 7×10^{-5} mol L^{-1}. The reference solution must be the same concentration of the surfactant and additive, if any. Record the spectrum of the sample. Repeat recording the spectrum for various concentrations (from 5.0×10^{-4} to 2.0×10^{-2}) of the surfactant.

Treatment of Data:

1. Determine the absorbance at the two wavelengths, namely 312 nm and 250 nm, corresponding to the enol form and the keto form respectively.
2. Plot the absorbances as a function of the surfactant concentration and note the point of gradient change.
3. The concentration of SLS corresponding to the point at which the slope changes abruptly gives the CMC value.

Result:

Report the value of the CMC of the surfactant.

Note:

Alternatively, the critical micelle concentration of surfactants can also be determined by measuring the fluorescence of an appropriate molecule in the surfactant medium or the electrical conductivity of the surfactant itself. These methods are discussed below.

FLUORESCENCE METHOD

Pyrene is one of the few aromatic hydrocarbons that show significant fine structure in the monomer fluorescence spectra in solution. The intensities of the vibrational bands of pyrene showed a strong dependence on solvent environment. In the absence of the surfactant, pyrene senses the polar environment of the water (solvent) and the ratio of fluorescence emission intensities corresponding to the first and third vibrational peaks (I_i/I_{iii}) is high. Above the CMC, owing to the high hydrophobicity of pyrene molecules these are solubilized in the interior of the micellar phase. This is a hydrocarbon like solvation situation and hence the polarity felt by pyrene is less and hence the ratio of (I_i/I_{iii}) decreases. If one were to plot (I_i/I_{iii}) as a function of the concentration of the surfactant it will show a sharp decrease at the micellar concentration. This could be utilized to determine the CMC of the surfactant.

Procedure: Prepare a pyrene solution by dissolving 5 mg in10 mL of methanol. Dilute the solution 20 times. A small aliquot (say 50 μL) of the methanolic pyrene solution is transferred to the quartz fluorescence cell and mixed with appropriate volumes of water, surfactant solution (0.03 mol L^{-1}) and additive if any to get a final volume of 3 mL. The concentration range of the surfactant (sodium lauryl sulphate, SLS) to be used is 1×10^{-3} to 12×10^{-3} mol dm^{-3}. The fluorescence intensity is measured at 373 and 384 nm after exciting with 334 nm radiation by the appropriate use of the emission slits.

Treatment of the Data:

1. Record the emission intensities corresponding to both the wavelengths.
2. Plot the relative intensity (I_i/I_{iii}) as a function of the concentration of the surfactant.
3. Find out the value of the concentration of the surfactant corresponding to the steep decrease in intensity.

Result:

Report on the value of the CMC.

ELECTRICAL CONDUCTIVITY METHOD

This method can be applied to the CMC of ionic surfactants. The change in the electrical conductance of aqueous ionic surfactant solutions at the CMC is due to different degree of surfactant ionization. Below CMC only the monomers are present and hence behave as a strong electrolyte and above CMC it behaves as partially ionized system. Therefore for the aqueous surfactant solutions, Kohlrausch's law is applicable and the specific conductivity, κ, can be calculated in terms of the molar ionic conductivities of ions $\lambda_i = z_i\mu_i F$, where z and μ are the charge and mobility of the ion and F is the Faraday constant.

Consider the Case SDS

Below the CMC

The specific conductivity of aqueous SDS solution in $mS.cm^{-1}$ is made up of independent contributions of anions $CH_3(CH_2)_{11} - OSO_3$, SD^- anions and Na^+ cations. Therefore,

$$\kappa = \left(\lambda_{Na^+} + \lambda_{SD^-}\right)[SDS]_t = m_1[SDS]_t \tag{1}$$

where m_1 is the slope of the plot of κ versus $[SDS]_t$ below the CMC.

Above CMC

The conductivity of ionic surfactants such as SDS decreases. This can be understood by the inclusion with the micelle of ions of opposite charge (counter ions) to the long-chain ions. The percentage of counter ions in relation to the number of long-chain ions in the micelle (the aggregation number N_{agg}) is called the fraction of micellar charge neutralized β; then α (= 1 – β) is the degree of micellar ionization. Therefore the concentration of free counter ions (Na^+ in the case of SDS surfactant) is given by

$$[Na^+]_f = CMC + \alpha\,[SDS]_m$$

where $$[SDS]_m = [SDS]_t - CMC$$

This essentially means the conductivity above CMC is contributed due to single ions (Na^+ and SD^-) at the CMC, due to micellar ions and that is due to the Na^+ ions in excess.

The specific conductivity of a micellar solution can be written as

$$\kappa = \left(\lambda_{Na^+} + \lambda_{SD^-}\right)CMC + \lambda_{Na} + \alpha[SDS]_m + \lambda_{mic}[\text{micelles}] \tag{2}$$

Since [micelles] = ($[SDS]_t$ – CMC)/N_{agg} and assuming the contribution of the micelle to the conductance is the same as that of an equivalent number of monomeric ions, the sum of those charges equals the micellar charge, then $\lambda_{mic} = \alpha N_{agg}\,\lambda_{SD}$. Therefore substituting this quantity in the above equation gives,

$$\kappa = \left(\lambda_{Na^+} + \lambda_{SD^-}\right)CMC(1-\alpha) + \left(\lambda_{Na} + \lambda_{SD}\right)\alpha[SDS]_t$$

$$= \kappa_0 + m_2 \,[SDS]_t \qquad (3)$$

where $m_2 = m_1 \alpha$ is the slope of the linear plot of κ versus $[SDS]_t$ above CMC and κ_0 is the corresponding intercept.

Therefore the CMC can be determined either as the cross point of the two straight lines defined by equations 1 and 3. Alternatively one can also obtain the value of CMC from the intercept of the plot according to equation 3.

Procedure: Prepare a solution of the surfactant (say 0.05 mol L^{-1}). Measure the specific conductivity as usual at each [surfactant] following the step by step dilution extraction method. The conductivity measurements have to be carried out in a jacketed beaker at 298 K by circulating water.

Treatment of Data:

1. Plot the value of the specific conductance as a function of the surfactant concentration.
2. Draw smooth lines and find the intersection point of two straight lines.

Result:

Report the value of the CMC.

PART VIII

Conductometry

EXPERIMENT 52

Determination of Cell Constant

Aim: To determine the cell constant of a conductivity cell.

Apparatus Required: Conductivity cell, conductivity bridge, 100 cm^3 volumetric flasks and pipette.

Chemicals Required: Potassium chloride AR quality, conductivity water.

Principle: A conductivity cell consists of two identical parallel platinum plates (electrodes) sealed into a glass tube. The assembly is show in (Fig. 27). The ratio of the length (l, distance between the two electrodes) to the cross section area (a) of the electrode is referred to as the cell constant. As the name indicates, this value is constant for a given cell. Its value depends on the resistance (R) offered by the electrode to the current flowing through the electrolytes and the specific conductance (κ) which is the conductance of the electrolyte kept in a conductivity cell with electrodes of 1 cm^2 area and 1 cm apart. The relation is given as,

$$\text{Cell constant} = R \times \kappa$$

The unit of cell constant is cm^{-1}. In SI system it is expressed as m^{-1}.

The resistance of the electrolyte solution is measured employing the Wheatstone's Bridge Principle.

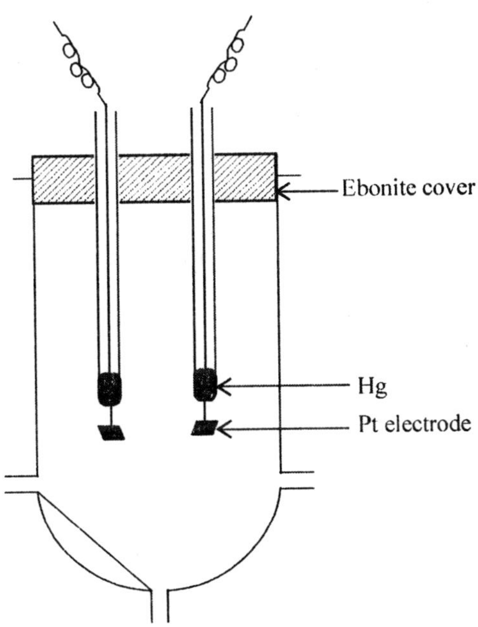

Fig. 27. Conductivty cell.

In practice, KCl solutions for which the specific conductance values are known are taken in the cell and from the measured resistance values, the cell constant may be obtained using the above relationship.

Procedure: Prepare 0.1 N and 0.01 N KCl solutions by accurately weighing the required quantity of the salt and making up to 100 cm^3 in volumetric flask, with conductivity water. Rinse the conductivity cell twice with conductivity water and then with the 0.01 N KCl solution. Take sufficient volume of 0.01 N KCl solution in the conductivity cell so that both the electrodes are completely immersed in the solution. Connect this to the conductivity bridge and measure the resistance. Wash the cell carefully and repeat the above procedure with 0.1 N KCl solution.

Treatment of Data: Record the resistances of the KCl solutions and calculate the cell constant as,

$$\textit{Cell constant} = \textit{Specific conductance} \times \textit{Resistance.}$$

[KCl], N	Resistance (R)	Cell constant, cm^{-1}
0.01	. . . . (R1)	
0.1	. . . . (R2)	
		Mean =

The specific conductance values for the solutions may be obtained from standard tables.

Result:

Report the mean value of the cell constant at the experimental temperature.

Note:

The conductivity, for a known strength of KCl solution at 25°C may be calculated from the empirical relation* given below:

$$\lambda = 1.4982 \times 10^{-2} - 2.9678 \times 10 - 4\sqrt{C} + 9.49 \times 10^{-6}C$$
$$(1 - 7.191 \times 10^{-3}\sqrt{C})\ ohm^{-1}m^2mol^{-1}$$

where C is expressed in mol dm^{-3}.

*Shedlovsky, J.Amer.Chem.Soc. **54**, (1932), 1411.

EXPERIMENT 53

Verification of Onsager's Equation and Determination of Equivalent Conductance at Infinite Dilution of Strong Electrolytes

Aim: To verify Onsager's equation of electrolytic conductance and to determine the equivalent conductance at infinite dilution of a strong electrolyte.

Apparatus Required: Conductivity cell, conductivity bridge, 100 or 50 cm^3. Volumetric flasks and pipettes.

Chemicals Required: 0.5 N NaCl solution in conductivity water.

Principle: The equivalent conductance (λ) of a strong uni-uni valent electrolyte in dilute aqueous solution is related to its concentration (C) by the Debye-Hückel-Onsager equation as,

$$\lambda = \lambda_0 - (A + B\lambda_0)\sqrt{C}$$

The equivalent conductance can be calculated from the specific conductance (κ) using the equation,

$$\lambda = 1000 \times \kappa C$$

where the concentration, C is expressed in gram equivalents per dm^3.

When the λ values are plotted against $\sqrt{C}$, a straight line with a negative slope will be obtained, in accordance with the Onsager's equation. The Y–intercept of this line would then give the λ_0, the equivalent conductance at infinite dilution for the electrolyte.

Procedure: Dilute the 0.5 N NaCl stock solution quantitatively with conductivity water, appropriately to get solutions of 0.1, 0.05, 0.02, 0.01, 0.005 and 0.001 N concentration. Using a conductivity cell of known cell constant*, measure the resistance (or conductance) for each one of these solutions. Wash the cell every time with distilled water, rinse it with conductivity water and then with the solution under investigation, before making final measurement.

Treatment of Data:

Cell constant = ... cm^{-1}

[NaCl]	C	Resistance (R) ohm	$\kappa = (1/R) \times$ cell const.	$\lambda = 1000 \times \kappa/C$
0.001				
0.005				
0.01				
0.02				
0.05				
0.10				

Plot the λ values against the $\sqrt{C}$ as shown in the graph (Fig. 28). Extrapolate and find the y-intercept value. This corresponds to λ_o, the equivalent conductance at infinite dilution.

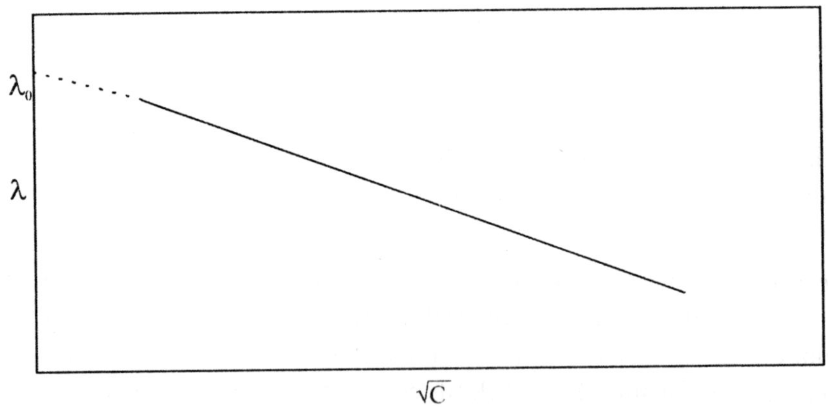

Fig. 28. Onsager's equation graph

Result:

Report the value for the equivalent conductance at infinite dilution, λ_o for NaCl in $ohm^{-1}cm^2eq^{-1}$.

*If it is not known, determine it as described in *experiment* 52.

EXPERIMENT 54

Verification of Ostwald's Dilution Law and Determination of Dissociation Constant of Weak Acids

Aim: To verify the Ostwald's dilution law for a given weak electrolyte and to determine its dissociation constant.

Apparatus Required: Conductivity bridge, conductivity cell, 100 cm^3 volumetric flasks and pipettes.

Chemicals Required: 0.1N acetic acid (or any weak electrolyte), conductivity water.

Principle: Weak electrolytes are only partially dissociated in solution. Hence, for such electrolytes the dissociation constant (K) is given by the Ostwald's dilution law as $K = C\alpha^2/(1 - \alpha)$ where C is its molar concentration, and α is the degree of dissociation. The value of α is given as the ratio of the equivalent conductivity of the electrolyte at a particular concentration to that at infinite dilution. i.e. $\alpha = \lambda/\lambda_0$. However, in such cases λ_0 may be determined only by the application of Kohlrausch's law of independent migration of ions.

Procedure: Prepare different concentrations, say 0.05, 0.02, 0.01, 0.005, 0.002 N of the given weak electrolyte by accurately diluting the given stock solution, in conductivity water. Using conductivity cell of known cell constant, measure the resistance for each of the above solutions. Calculate the specific conductivity and using the λ_0 value for the electrolyte compute the dissociation constant, at the experimental temperature.

Treatment of Data:

Cell constant (l/a) of the cell = ...cm^{-1}

λ_o for the given electrolyte = $ohm^{-1}\ cm^2\ eq^{-1}$

Compute the specific conductance and equivalent conductance using the formulae

Specific conductance, $\kappa = (1/R) \times$ cell constant, and

Equivalent conductance, $\lambda = 1000 \times \kappa/C$.

Conc. (C) N	Resistance (Ohm) R	Sp. cond. (ohm^{-1}) κ	Eq. cond. λ	$\propto = \lambda/\lambda o$	$K = C\alpha^2/(1-\alpha)$
0.002					
0.005					
0.010					
0.020					
0.050					
0.100					

Mean dissociation constant = . . .

As the degree of dissociation , $\propto$, is very small for a weak electrolyte, $\propto << 1$ and hence the Ostwald's dilution law may be written as,

$$K = C\propto^2$$

or

$$\propto = \sqrt{(K/C)}$$

Therefore a plot of $\propto$ versus $1/\sqrt{C}$ is expected to be linear. The slope of the line will give $\sqrt{K}$, from which K can be found out.

Result:

Report the dissociation constant value for the given weak electrolyte evaluated by

(i) Calculation and (ii) Graphical method.

EXPERIMENT 55

Verification of Walden's Rule

Aim: To verify the Walden's Rule, $\lambda_0\eta_o$ = constant.

Apparatus Required: Conductivity cell, conductivity bridge, 100 cm^3 volumetric flasks and pipettes.

Chemicals Required: N-hexadecyl – N, N, N – trimethylammonium bromide, conductivity water, methanol, and acetone.

Principle: Walden's Rule may be stated as the product of the equivalent conductance at infinite dilution (λ_0) of an electrolyte in a given solvent and the viscosity (η_0) of that solvent is a constant and independent of the nature of the solvent, (i.e.) $\lambda_0\eta_o$ = constant.

This relationship may be expected from the Stoke's Law: $f = 6\pi\eta ru$, where a particle of radius, r moves with a steady velocity u, through a medium of viscosity η, on the application of a force, f. For a particular ion of radius r, the conductance (λ) is proportional to the mobility (velocity) of the ion. Hence, the Stoke's Law may be given as

$$\eta u = f/6\,\pi r$$

or

$$\eta\lambda = f/6\pi r$$

$$= \text{constant for a given ion moving under constant force.}$$

Procedure: Weigh accurately 0.396 g of N-hexadecyl-N, N, N-trimethylammonium bromide; dissolve it in conductivity water and make it up to 100 cm^3 in a volumetric flask. This corresponds to 0.01 M solution. Dilute this appropriately to get various lower concentrations say 0.005, 0.002, 0.001 M. Measure the conductivity of each of these solutions using a conductivity cell of known cell constant.

Then prepare the same concentrations of the compound in methanol and also in acetone. Determine the conductivity of each of these solutions also.

Treatment of Data:

Cell constant of the conductivity cell = ...cm^{-1}

Table 1. Conductance of the electrolyte solutions in different solvents.

[Electrolyte] M	Conductance, ohm^{-1} (1/R)			Sp. conductance κ = 1/R × cell const.			Eq. conductance λ = 1000 κ/c		
	Water	Methanol	Acetone	Water	methanol	acetone	Water	Methanol	Acetone
0.001									
0.002									
0.005									
0.01									

Plot the values of λ in each solvent (from the above Table 1) against $\sqrt{C}$; Extrapolate and find the y-intercept, which corresponds to λ_o, the equivalent conductance at infinite dilution.

Table 2. λ_0 values of the electrolyte in different solvents used.

Solvent	λ_0
Water	
Methanol	
Acetone	

Collect the viscosity values for these three solvents at the experimental temperature from reference tables.

Table 3. Viscosity of the solvents used.

Solvent	λ_0	η_o (viscosity)	$\lambda_o\eta_o$
Water			
Methanol			
Acetone			

Result:

Report the $\lambda_o\eta_o$ values in the different solvents.

EXPERIMENT 56
Conductometric Determination of Critical Micelle Concentration

Aim: To determine the critical micelle concentration of a surfactant by conductivity method.

Apparatus Required: Conductivity cell, conductivity bridge, 100 cm^3 volumetric flasks and pipettes.

Chemicals Required: Sodium lauryl sulphate (SLS), Conductivity water.

Principle: Surfactants are the molecules which get adsorbed at an interface and thus alters the surface tension/interfacial tension. Above a certain concentration, these molecules aggregate to form micelles, (ref. expt. 47), and this concentration is known as critical micelle concentration, (cmc). As micelles are the aggregate of individual molecules, their mobility is lower than that of single molecules. Consequently, micelles will have lower conductance. For further details with regard to the principles of this method refer to the note section in *Experiment 52.*

Procedure: Prepare 0.1 M solution of sodium lauryl sulphate by weighing accurately 2.88 g of it and dissolving in 100 cm^3 conductivity water in a volumetric flask. Dilute this solution appropriately to get various lower concentrations, say 0.05, 0.02, 0.01, 0.008, 0.006, 0.005, 0.004, 0.002 and 0.001 M. (Note: Solution should be shaken slowly and carefully to avoid excessive foaming). Measure the conductance of each of these solutions using a conductivity cell of known cell constant. From these values calculate the equivalent conductance (λ).

Treatment of Data:

Cell constant = ...cm^{-1}

[SLS] M	Conductance (ohm^{-1}) (1/R)	Sp. conductance κ = (1/R) × Cell const.	Eq. conductance $\lambda = 1000\ \kappa/C$
0.001			
0.002			
0.004			
0.005			
0.006			
0.008			
0.01			
0.02			
0.05			

Plot the values of equivalent conductance (λ) against the concentration (C) of SLS.

A graph consisting of two straight lines, as shown in (Figure 29), will be obtained.

Extrapolate the lines and find the point of intersection. The concentration corresponding to this point of intersection gives the critical micelle concentration of SLS.

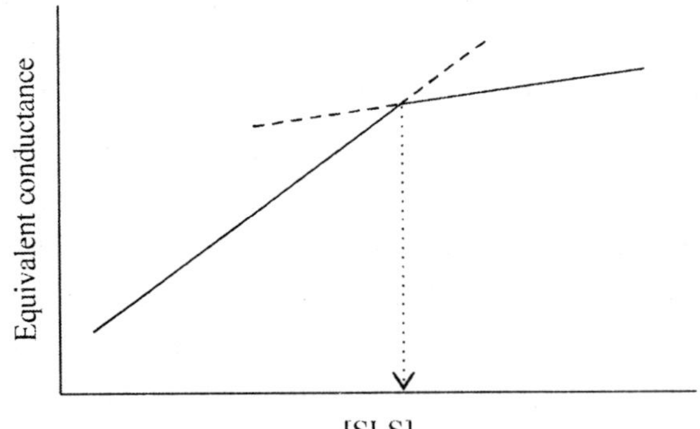

Fig. 29. Critical micelle concentration of SLS.

Result:

Report the critical micelle concentration value of the surfactant, sodium lauryl sulphate at the experimental temperature.

* For details see Experiment 51

EXPERIMENT 57
Titration of a Strong Acid with a Strong Base

Aim: To determine the concentration of a given strong acid by conductometric titration.

Apparatus Required: Conductivity bridge, conductivity cell, burette and pipettes.

Chemicals Required: Hydrochloric acid (in the range of 0.01M), sodium hydroxide (0.1M) — both prepared in conductivity water.

Principle: When a strong acid, say HCl, is titrated with a strong base, NaOH, the following reaction takes place

$$(H^{+} + Cl^{-}) + (Na^{+} + OH^{-}) \longrightarrow Na^{+} + Cl^{-} + H_2O$$

Thus, the highly conducting H^{+} ions initially present in the solution are replaced by Na^{+} ions having lower conductivity. Consequently, during the course of the reaction the conductance will be progressively decreasing. However, when excess alkali is added, due to the presence of more labile OH^{-} ions, the conductance will start increasing. Thus, at the equivalence point the conductance will be minimum.

Procedure: Pipette out 20 cm^3 of the given HCl solution into a clean conductivity cell of known cell const. Take the standardized solution of NaOH in a clean burette (a microburette is preferable). The strength of NaOH should be at least five times higher than that of the acid solution, so that the volume change would be minimal. Add the NaOH solution from the burette carefully, say just 0.1 cm^3 each time. After each addition note down the resistance and calculate the conductance value. Continue the titration, till the two limbs of the curve are clearly traced.

Treatment of Data:

Cell constant = ...cm^{-1}
Volume of HCl = ...cm^3
Strength of NaOH = ...N (N_{NaOH})

Volume of NaOH added (cm^3)	Resistance (ohm) (R)	Conductance (ohm^{-1}) (I/R)
0.1		
0.2		
0.3		
—		
—		
—		
—		
—		
—		

Plot the conductance value against the corresponding volume of NaOH added. A curve with two limbs as shown in Fig.30 will be obtained. The point at which these two lines intersect corresponds to the volume of NaOH (V_{NaOH}) required for exact neutralisation of the acid.

$$\therefore \text{ Strength of the given HCl } = \frac{V_{NaOH} \times N_{NaOH}}{20}$$

$$= \text{.............} \text{ N}$$

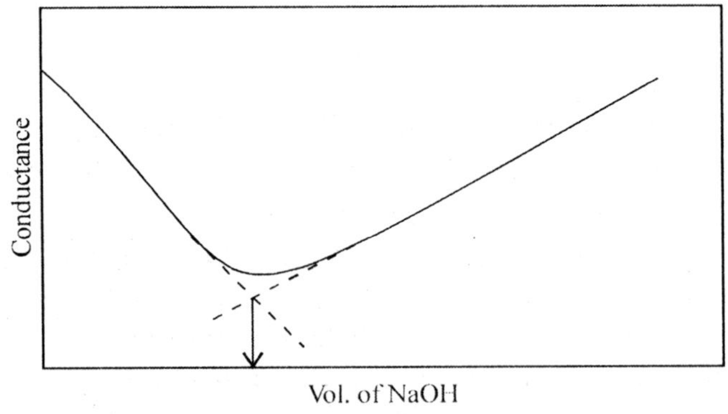

Fig. 30. Strong acid-strong base conductometric titration graph.

Results:

(i) Volume of NaOH at equivalence point, V_{NaOH} = ...cm^3.

(ii) Strength of the given strong acid = ...N.

EXPERIMENT 58
Titration of a Weak Acid with a Strong Base

Aim: To determine the concentration of a given weak acid by conductometric titration.

Apparatus Required: Conductivity bridge, conductivity cell, burette and pipettes.

Chemicals Required: Acetic acid (in the range of 0.01M); standard NaOH solution (in the range of 0.1M); both prepared in conductivity water.

Principle: When a moderately weak acid, say acetic acid, is titrated with a strong base like NaOH the following reaction takes place.

$$CH_3COOH + NaOH \rightarrow CH_3COO^-Na^+ + H_2O$$

Acetic acid, being weak, ionizes only partially. So, the conductance for the solution, initially will be low. On adding a very small amount of alkali, CH_3COONa will be formed. However, the common ion (acetate ion) will suppress the ionization of acetic acid; hence, there may be a further decrease in the conductance initially. However, this initial suppression is overcome at higher alkali concentrations and the conductance starts progressively increasing. When the acid is completely neutralized, further addition of alkali will result in the presence of highly conducting Na^+ and OH^- ions in the solution, associated with a steep increase in conductance. This break point will give the equivalence point.

Procedure: Pipette out 20 cm^3 of the given acetic acid solution into a clean conductivity cell of known cell constant. Add the standardized NaOH solution taken in a burette (preferably in a microburette) in small aliquots, say 0.1 cm^3, to the acid. After each addition, stir the contents well and note down the resistance or conductance value. Continue the titration till the two limbs of the curve (Fig. 31) are clearly traced.

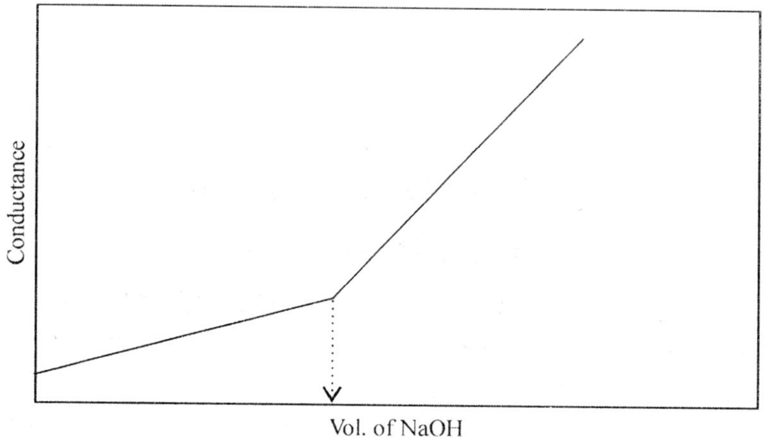

Fig. 31. Weak acid—strong base conductometric titration graph.

Treatment of Data:

Cell constant = ...cm^{-1}

Volume of the weak acid = 20 cm^3

Strength of NaOH = ...N (N_{NaOH})

Volume of NaOH Added (cm^3)	Resistance (ohm) (R)	Conductance (ohm^{-1}) (I/R)
0.1		
0.2		
0.3		
—		
—		
—		
—		
—		

Plot the conductance value against the corresponding volume of NaOH added. A curve with two limbs as shown in the (Fig. 31) will be obtained. The point of intersection of these two lines corresponds to the volume of NaOH (V_{NaOH}) required for exact neutralization of the acid.

$$\therefore \text{ Strength of the given weak acid } = \frac{V_{NaOH} \times N_{NaOH}}{20}$$

$$= \text{..............} \text{ N.}$$

Results:

(i) Volume of NaOH required at equivalence point = ...cm^3.

(ii) Strength of the given weak acid = ...N.

EXPERIMENT 59
Titration of a Mixture of Strong and Weak Acid with a Strong Base

Aim: To determine the concentration of a strong acid and a weak acid present in a mixture.

Apparatus Required: Conductivity bridge, conductivity cell, 100 cm^3 beaker, 100 cm^3 volumetric flasks, burette and pipettes.

Chemicals Required: Standardised NaOH solution (0.02 N), HCl solution (0.001N), acetic acid (0.01N) solution.

Principle: Upon adding a strong base to a mixture of a strong acid (HCl) and weak acid (CH_3COOH), the following reactions occur sequentially.

$$HCl + NaOH \longrightarrow NaCl + H_2O$$

$$CH_3COOH + NaOH \longrightarrow CH_3COONa + H_2O$$

First, the conductance falls on adding the strong base to the mixture until the strong acid is fully neutralized; then rises as the weak acid is converted to its salt, a strong electrolyte and finally rises more steeply as the alkali is introduced in excess.

Procedure: Pipette out 10 cm^3 of the given HCl solution and 5 cm^3 of acetic acid solution into a clean conductivity cell of known cell constant. Fill the burette with the standard NaOH solution (a micro-burette is preferable). A small aliquot (say 01 cm^3) of NaOH solution is added to the acid mixture; the solution is stirred well with a glass rod and the conductance noted. Continue the addition of the base and measure the concentration until the three limbs of the graph (Fig. 32) are traced fully.

Treatment of Data:

Cell constant = ...cm^{-1}

Strength of NaOH = ...N (N_{NaOH})

Volume of HCl = 10 cm^3

Volume of acetic acid = 5 cm^3

Volume of NaOH Added (cm^3)	Resistance (ohm) (R)	Conductance (ohm^{-1}) (1/R)
0.1		
0.2		
0.3		
—		
—		
—		
—		
—		

Plot the conductance value against the corresponding volume of NaOH added. A curve with three limbs as shown in (Fig. 32), will be obtained. The point A corresponds to the neutralization of the strong acid; hence the volume of NaOH corresponding to the point A is the volume required for the neutralization of HCl (V_A). Similarly the point B corresponds to the volume of NaOH required to neutralize all the acids. Hence, volume corresponding to (B – A) gives the NaOH required to neutralize the weak acid, (V_{B-A}).

$$\therefore \quad \text{Strength of the given strong acid} = \frac{N_{NaOH} \times V_A}{15}$$
$$= \text{.................. N.}$$

$$\text{Strength of the given weak acid} = \frac{N_{NaOH} \times V_{(B-A)}}{15}$$
$$= \text{.................. N.}$$

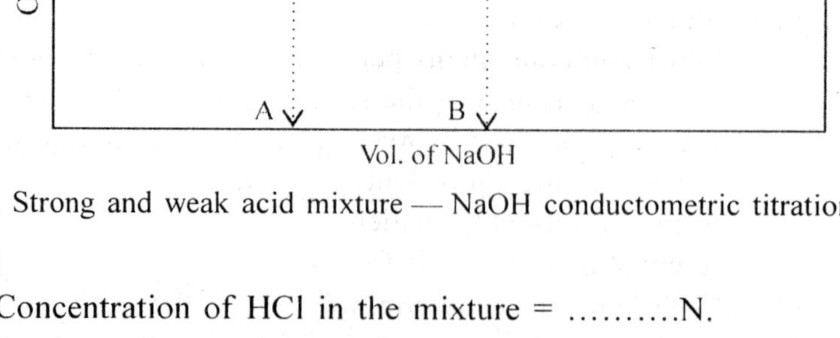

Fig. 32. Strong and weak acid mixture — NaOH conductometric titration graph.

Results:

(i) Concentration of HCl in the mixture = ...N.

(ii) Concentration of CH_3COOH in the mixture = ...N.

EXPERIMENT 60

Precipitation Titration and Determination of Solubility of a Sparingly Soluble Salt

Aim:

1. To follow conductometrically the titration of an aqueous solution of lead nitrate against sodium sulphate solution and
2. To determine the solubility of lead sulphate.

Apparatus Required: Conductivity bridge, conductivity cell, 100 cm^3 beakers, micro-burette (10 cm^3) and pipettes.

Chemicals Required: Standard sodium sulphate and lead nitrate solutions and the sparingly soluble salt (lead sulphate).

Principle: Sodium sulphate and lead nitrate are water soluble salts. However, when they are titrated one against the other, the product lead sulphate formed is insoluble and get precipitated.

$$Na_2SO_4 + Pb(NO_3)_2 \longrightarrow PbSO_4 \downarrow + 2\,NaNO_3$$

Where a precipitate is formed, one salt is replaced by an equivalent amount of another. E.g., sodium sulphate by sodium nitrate and so the conductance remains almost constant in the early stages of the titration. After the equivalence point is passed, however, the excess of the added salt causes a sharp rise in conductance (See Fig. 33), the end point of the reaction can thus be determined.

If 's' is the solubility in equivalents per liter of a given substance and κ (kappa) is the specific conductance of the saturated solution, the equivalent conductance of the solution is given by $\lambda = 1000\ \kappa/s$. In general the solution will be sufficiently dilute for the equivalent conductance to be little different from the value at infinite dilution(λ_0) which can be obtained from the ionic conductances of the constituent ions. It follows that since λ is known and κ for the saturated solution can be determined experimentally, it is possible to evaluate the solubility 's' by means of the above equation.

Procedure:

1. Conductometric titration of lead nitrate versus sodium sulphate solution:

Determine the cell constant of the given conductivity cell. Pipette out 50 cm^3 of the lead nitrate solution into a 100 cm^3 beaker and determine the initial conductance of the solution. Determine the conductance after each addition (0.2 cm^3) of standard sodium sulphate solution from the burette. Solution should be stirred well after each addition. Plot the conductance of the solution against the volume of the sodium sulphate solution added (Fig. 33). Determine the end point from the graph and calculate the strength of the lead nitrate solution.

2. ***Determination of solubility:*** Wash the precipitate of lead sulphate that had formed in the beaker thoroughly several times with conductivity water and then find out the conductance of the saturated solution of lead sulphate.

Calculate solubility from the formula, solubility = 1000 κ/λ_0 where κ is the specific conductance of the precipitated salt, and λ_0, equivalent conductance at infinite dilution.

Tabulation and Calculation

1. Determination of the strength of $Pb(NO_3)_2$ solution

Weight of $Pb(NO_3)_2$ =

Normality of lead nitrate solution =

Cell constant of the cell used =

Normality of sodium sulphate solution = . . . N

Volume of lead nitrate solution taken = . . . cm^3

Volume of Na_2SO_4 solution added										
Conductance Ohm^{-1}										

Volume of sodium sulphate (from the graph) = . . . cm^3

Normality of lead nitrate solution = . . . N

2. Determination of solubility:

Resistance of saturated solution of lead sulphate = ohm

λ_0 of lead sulphate = 149.5 $ohm^{-1}\ cm^2\ g\ eq^{-1}$

κ of saturated solution of lead sulphate = $ohm^{-1}\ cm^{-1}$
Specific conductance of conductivity water = $ohm^{-1}\ cm^{-1}$
Specific conductance of lead sulphate = $ohm^{-1}\ cm^{-1}$
Solubility of lead sulphate = g eqv./litre.

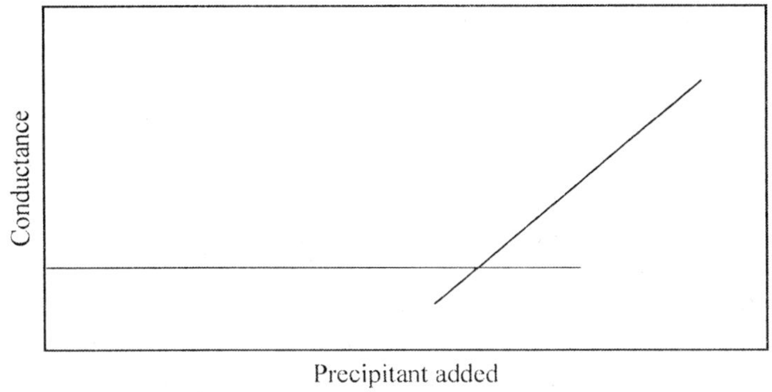

Fig. 33. Precipitation titration graph.

Results:

(i) Strength of the given lead nitrate solution = ...N

(ii) Solubility of lead sulphate = ...g/L

Note:

1. Better accuracy in solubility can be obtained by substituting κ value as the difference in κ values determined for the saturated solution of the sparingly soluble salt and that for pure conductivity water.
2. Solubility in eq L^{-1} can be converted to g L^{-1} by multiplying the value in eq L^{-1} with weight of the salt.
3. If both the products of the reaction are sparingly soluble, as for example in the titration of magnesium sulphate by barium hydroxide

$$(Mg^{2+} + SO_4^{2-}) + (Ba^{2+} + 2\ OH^-) = Mg(OH)_2 + BaSO_4$$

the conductance of the solution decreases right from the commencement, but increases after the end point because of the free barium hydroxide. Typical plots will result as shown in Fig 34.
4. Precipitation reactions cannot be carried out conductometrically with great accuracy due to the slow separation of the precipitation with consequent super saturation of the solution and removal of titrated solute by adsorption on the precipitate etc. The best results have

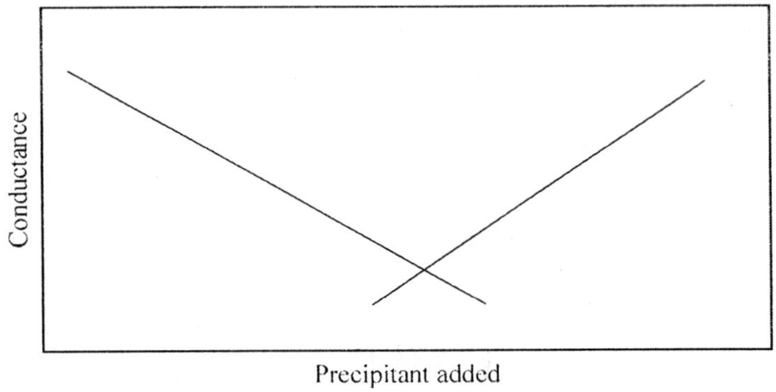

Fig. 34. Precipitation titration graph.

been obtained by working with dilute solutions, in the presence of a relatively large amount of alcohol, the latter causes a diminution of the precipitate and there is also less adsorption.

PART IX
Potentiometry

EXPERIMENT 61

Transport Numbers of Silver and Nitrate Ions by Hittorf's Method

Aim: To determine the transport number of silver and nitrate ions by Hittorf's method.

Apparatus Required: Hittorf's apparatus, silver electrodes, rectified ammeter, plastic wires and standard flasks.

Chemicals Required: 0.1N silver nitrate, 0.1N NH_4CNS and ferric alum indicator.

Principle: The transport (or transference) number of cation and anion is represented as t_+ and t_- respectively and are given by

$$t_+ = u_+/(u_+ + u_-) \text{ and } t_- = u_-/(u_+ + u_-)$$

where u_+ and u_- are the mobilities of the respective ions. $t_+ + t_- = 1$.

The Hittorf's method of determining transference numbers involves measurement of changes of concentration in the vicinity of the electrodes.

Suppose an electric current is passed through the solution of an electrolyte which yields the ions M^+ and A^- (these ions are not necessarily univalent). The fraction of the total current carried by the cations is t_+ and that carried by the anions is t_- hence when one Faraday of electricity is passed through the solution, t_+ Faradays are carried in one direction by t_+ equivalents of M^+ ions and t_- Faradays are carried in the other direction by t_- equivalents of A^- ions. At the same time one equivalent of each ion is discharged at the appropriate electrode. The migration of the ions and their discharge under the influence of the current bring about changes of concentration in the vicinity of the electrodes and from these changes it is possible to calculate the transference numbers.

Imagine the cell containing the electrolyte to be divided into three compartments by means of two hypothetical partitions one compartment surrounds the cathode, another the anode and the third is a middle compartment in which there is no resultant change in concentration. The effect of passing one Faraday of electricity through the solution of the electrolyte *MA* can then be represented in the following manner.

Cathode compartment (I)	Middle compartment	Anode compartment (II)
1 equivalent of M^+ is discharged	t_+ equivalents of M^+ migrated to I	1 equivalent of A^- is discharged
t_+ equivalent of M^+ migrate in	t_- equivalents of A^- migrate from I	t_- equivalent of A^- migrate in
t_- equivalent of A^- migrate out	t_+ equivalent of M^+ migrate from II	t_+ equivalent of M^+ migrate out
Net Result	t_- equivalent of A^-	Loss of $1 - t = t_+$
Loss of $1 - t_+ = t$	migrate to II	Equivalent of A^-
Equivalents of M^+	Net Result	Loss of t_+
Loss of t_-	No change in concentration	Equivalent of M^+
Equivalents of A^-		Net Loss is t_+
Net Loss of t_-		Equivalent of MA
Equivalents of MA		

It follows therefore, if the discharged ion may be regarded as being completely removed from the system and the electrodes are not attacked as is tacitly assumed in the above tabulation, that (equivalent of electrolyte lost from the anode compartment) / (equivalent of electrolyte lost from the cathode compartment) = t_+/t_-.

The total decrease in amount of electrolyte MA in both compartments of the experimental cell is equal to the number of equivalents deposited on each electrode, if a coulometer is included in the circuit, then by Faraday's laws the same number of equivalents of material, no matter what its nature will be deposited. It follows therefore, that equivalent of electrolyte lost from anode compartment/equivalent deposited on each electrode of cell or in coulometer = t_+ and equivalent of electrolyte lost from cathode compartment/ equivalent deposited on each electrode of cell or in coulometer = t_-.

By measuring the fall in concentration of electrolyte in the vicinity of the anode and cathode of an electrolytic cell and at the same time determining the amount of material deposited on the cathode of the cell or the coulometer in circuit it is possible to evaluate the transference numbers of ions present in solution.

Procedure: Fill the apparatus with 0.1N silver nitrate, taking care to avoid the presence of air bubbles. Connect one silver electrode to the positive terminal and the other to the negative terminal of the rectifier through an ammeter. Connect the rectifier to the mains. Pass a steady current of 7.5 mA through the solution for about 2-3 hours. Drain the solution in each of the compartments into previously weighed conical flasks and weigh the flasks once again with the solutions. Titrate the contents of the flask against standardized NH_4CNS using ferric alum as the indicator.

Treatment of Results:

Consider the process taking place at the cathode.

1. Find out the initial concentration of Ag in the solution (c g per 1000 g of water).
2. Calculate the number of coulombs passed through the apparatus 'Q' and hence the amount of Ag deposited at the electrode.

   $Q \times 107.88/96500 = N$ g where Q = current × time.
3. Find out the amount of Ag after electrolysis.
4. Express this amount of Ag in grams per weight of water (y grams in (w – y) grams of water, where w is the weight of the solution).
5. Determine the amount of Ag in (w – y) grams of water from the initial concentration [$c \times (w - y)/1000$ grams of Ag].
6. Determine the amount of Ag lost in the catholyte

   [$y - (c \times (w - y)/1000) = M$ g.
7. Find out the amount that has migrated away $(N - M)$ g.
8. The fraction of the current carried by the amount $(N - M)/N$ = transport number of $Ag^+ = t_{Ag^+}$;
9. Calculate $t_{NO_3^-} = (1 - A_{Ag^+}) = t_{NO_3^-}$

   Perform similar calculations for the anolyte also.

Weight of the bottle + $AgNO_3$ .... = g

Weight of bottle .... = g

Weight of $AgNO_3$ = . . . g

Normality of $AgNO_3$ = . . . N

Standardisation of NH_4CNS =

Volume (cm^3) of $AgNO_3$ = Volume (cm^3) of NH_4CNS

$\therefore$ Strength of NH_4CNS = . . . N

Current passed = . . .milliamp

Time = . . . min.

Initial concentration of $AgNO_3$ = . . . g

Weight of water associated with ...... g of Ag^+ = . . . g

Weight of Ag^+ per 1000 g of water = . . . g

Amount of Ag dissolved at anode or deposited on the cathode =

Analysis of the cathode solution

Weight of the flask + cathode solution = . . . g

Weight of flask = . . . g

Weight of solution = . . . g

Volume of NH_4CNS required for the catholyte. = . . . cm^3

Therefore, amount of silver in the catholyte after electrolysis = . . . g

Weight of water = . . . g

Amount of silver initially associated with ...... g of water =

Loss of Ag^+ in the catholyte = . . . g

Amount of Ag^+ which has migrated = . . . g

Transport number of Ag^+ =

Transport number of NO_3^- =

Analysis of the anode solution:

Weight of the flask + anolyte = . . . g

Weight of the flask = . . . g

Weight of anolyte	= . . . g
Volume of NH_4CNS required for the anolyte	= . . . cm^3
Amount of Ag in the anolyte after electrolysis	= . . . g
Weight of water	= . . . g
Amount of Ag initially associated with ...... g of water	= . . . g
Gain in Ag^+ in the anolyte	= . . . g
Transport number of Ag^+	=
Transport number of NO_3^- ion	=
Mean transport number of Ag^+ ion	=
Mean transport number of NO_3^- ion	=

Result:

Report the transport number of Ag^+ and NO_3^-.

EXPERIMENT 62

Standard Electrode Potential of Cu and Ag Electrodes and Determination of the Potential Difference in a Concentration Cell

Aim: To determine the standard electrode potentials of Cu and Ag electrodes and to determine the EMF of a concentration cell.

Apparatus Required: Potentiometric assembly, standard cell, calomel electrode, salt bridge, Cu and Ag electrodes, plastic wires, 100 cm^3 beakers and volumetric flasks.

Chemicals Required: 0.1M, 0.01M and 0.001M copper sulphate and silver nitrate solutions.

Principle: The principle generally employed in the measurement of the EMF of voltaic cells is that embodied in the Poggendorff compensation method. It has the advantage of giving the EMF of the cell on open circuit i.e. when it is producing no current. A cell can be expected to behave reversibly only when it is producing an infinitesimally small current and hence the condition of open circuit is the ideal one for determining the reversible EMF. Normal potentiometer assembly is shown in (Fig. 35).

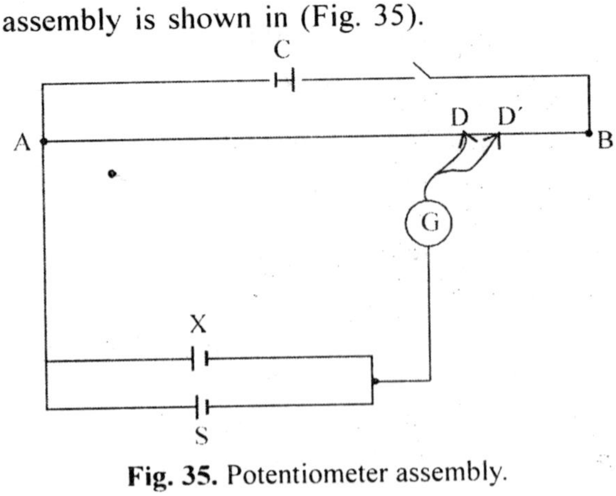

Fig. 35. Potentiometer assembly.

It consists of a working cell C, generally a storage battery of constant EMF which must be larger than that of the cell to be measured, connected across the ends of a uniform conductor AB of high resistance. The cell X, which is being studied is connected to A with the poles in the same direction as the cell C, and then through a galvanometer on to a sliding contact D which can be moved along AB. The position of D is adjusted until no current flows through the galvanometer, the fall of potential between A and D due to the cell C is exactly compensated by EMF of X, by means of a suitable switch, the cell X is replaced by a standard cell S, say a Weston Cadmium cell (see the note below) whose potential is 1.0183 V. The position of D is adjusted such that no current flows through the galvanometer.

Then the potential drop between A and D is exactly 1.0183 V and the voltage drop anywhere along A and B is identical with the voltage markings on the slide wire. The potentiometer has now been standardized to read the voltage directly.

Single electrode and cell potentials are determined not only by the nature of the constituents composing the electrodes, but also by the temperature and the activities of the solutions employed. The dependence of the EMFs on the latter variables is deducible from thermodynamics. For any reaction such as

$$aA + bB + \text{..........} \rightleftharpoons cC + dD + \text{..........}$$

The change in free energy, ΔF as a function of the initial activities of the reactants and the final activities of the products is given by the reaction isotherm as

$$\Delta F = \Delta F^0 + RT \ln \{a_C{}^c a_D{}^d \ldots / a_A{}^a a_B{}^b \ldots\}$$

where ΔF^0 is the standard free energy change which is the free energy change attending the reaction when the activities of products and reactants are all unity.

Since $\Delta F = - nFE$ and $\Delta F^0 = - nFE^0$ the equation becomes

$$E = E^0 - (2.303RT/nF) \log_{10}(a_C{}^c a_D{}^d \ldots / a_A{}^a a_B{}^b \ldots)$$

R = 8.314 Joules, F = 96500 Faradays, T = 298 K

$$E = E^0 - (0.0591/n) \log_{10} (a_C{}^c a_D{}^d \ldots / a_A{}^a a_B{}^b \ldots)$$

E^0 is the EMF of the electrode when all the activities are unity.

$$Cu^{2+} + 2e^- \rightleftharpoons Cu$$

$$2Hg(l) + 2Cl^- \rightleftharpoons Hg_2Cl_2(s) + 2e^- \quad E^0 = -0.2415 = E_2$$

$$Cu^{2+} + 2\,Hg\,(l) + 2\,Cl^- \rightleftharpoons Hg_2Cl_2(s) + Cu$$

Hence $E_{obs} = E_1 + E_2$. Therefore E_1 can be calculated and from the above equation E^0 for Cu electrode can also be computed. Similar procedure can be followed for the silver electrode.

Procedure: First standardize the given potentiometer. For determining the standard electrode potential, set up a galvanic cell by dipping a Cu electrode into 0.1M solution of copper sulphate kept in a beaker and combining this with the reference (calomel) electrode through a salt bridge. Connect the positive end of the potentiometer to the copper electrode and the negative end of the potentiometer to the calomel electrode. Adjust the voltage setter together with the sliding contact until null deflection is obtained in the galvanometer and note the potential reading. Measure the EMF of the cell using different concentrations of copper sulphate solutions. Determine the EMF of a reversible cell (concentration cell) set up the cell by combining two half cells of copper electrodes in copper sulphate solutions of different concentrations through a salt bridge. Connect the positive terminal of the potentiometer to the copper electrode dipped in the more concentrated solution of copper sulphate and the negative terminal of the potentiometer to the other copper electrode dipped in the less concentrated solution. Measure the EMF. Repeat the above procedure for different concentrations in the half cells. Repeat the whole experiment with silver electrode and different concentrations of silver nitrate solutions.

Treatment of Data:

1. Calculate the standard electrode reduction potentials of Cu and Ag electrodes knowing the measured EMF, molarity and activity coefficient of copper sulphate and silver nitrate solutions using the relations

$$E_{observed} = E_1 + E_2$$

   where E_2 = EMF of the saturated calomel electrode (– 0.2415 V) and $E_1 = E^0 + (0.0591/n) \log^{a}_{cu^{2+} \text{ or } Ag+}$ where a = activity.
2. If E_1 is the potential of the more concentrated half cell and E_2 is the potential of the less concentrated half cell, then $E_1 - E_2$ = EMF of the concentration cell.

I. Determination of the standard electrode potential of Cu electrode

Weight of copper sulphate = . . . g

Normality of copper sulphate solution = . . . N

$[CuSO_4 \cdot 5H_2O]$ M	$F_{Cu^{++}}$	E_{obs} volt	$E_1 = E_{obs} + 0.2415$ volt	$E^0_{cu/cu^{++}} = E_1 - (0.0591/2) \log a_{cu^{++}}$ volt

Mean = ... volts

II. Determination of EMF of concentration cells

C_1	C_2	E_{obs} volt	$E_1 = E_2$ volt

III. Determination of standard electrode potential of Ag

Weight of $AgNO_3$ = . . . g

Strength of the silver nitrate solution = . . . N

$[AgNO_3]$	F_{Ag^+}	E_{obs} volt	$E_1 = E_{obs} + 0.2415$ volt	$E^0_{Ag/Ag^+} = E_1 - (0.0591/2) \log a_{Ag^+}$ volt

Mean = ... volts

IV. Determination of EMF of concentration cells

C_1	C_2	E_{obs} volt	$E_1 = E_2$ volt

Result: Report the standard reduction potential of copper and silver electrodes.

Note:

1. Standard Cell: One of the electrodes of the Weston cadmium cell is a 12.5 % cadmium amalgam in a saturated solution of $3CdSO_4 \cdot 8H_2O$ and the other electrode consists of mercury and solid mercurous sulphate in the same solution thus

   12.5% Cd in Hg/$3CdSO_4.8H_2O$ saturated solution, Hg_2SO_4(s)/Hg

   The cell is set up in a H-shaped tube, the left hand limb containing the cadmium amalgam and the right hand, the mercury; the amalgam is covered with crystals of $3CdSO_4 \cdot 8H_2O$ and the mercury with solid mercurous sulphate and the whole cell is filled with a saturated solution of cadmium sulphate. The EMF of Weston standard cell in volts, over a range of temperatures is given by the expression

   $$E_t = 1.018300 - 4.06 \times 10^{-5}\,(t - 20) - 9.5 \times 10^{-7}\,(t - 20)^2 + 1 \times 10^{-8}\,(t - 20)^3$$

   so that the value is 1.0183 volt at 20°C and decreases about 4×10^{-3} millivolt per degree in this region.

   Report the EMFs of the concentration cells used and show how they compare with the calculated values.

2. Concentration Cells: Concentration cells may be made up of two electrodes of the same metal, or other material, at different concentrations immersed in solutions containing ions of the metal. Suppose the activities of the metal ions in the two solutions of concentrations c_1 and c_2 are a_1 and a_2 then the removal of 1 mole of metal ions from the more concentrated solution (c_1) and its deposition in the dilute solution (c_2) requires mzF coulombs where m is the number of atoms in the molecule and z is the valency of the ion. The electrical work equal to $-\Delta F$ is thus mzFE where E is the EMF of the cell. The free energy change accompanying the transfer of 1 mole of metal ion from activity a_1 to a_2 is

   $$\Delta G = \mu_2 - \mu_1 = RT \ln a_2/a_1$$

   at constant temperature and pressure; hence $E = (RT/mzF) \ln a_1/a_2$ or the EMF of the concentration cells will be the differences in EMF of the half cells comprising it.

EXPERIMENT 63
Thermodynamic Parameters of a Reaction from EMF Measurement

Aim: To determine the thermodynamic parameters for a reaction from EMF measurements.

Apparatus Required: Potentiometer assembly, H shaped cell with zinc amalgam and silver chloride solution.

Chemicals Required: 0.1M $ZnCl_2$, 0.001M HCl.

Principle: The net electrical work performed by a reaction yielding an EMF and supplying a quantity of electricity Q is $Q \times E$. For each equivalent reacting, Q is equal to the Faraday F and hence for n equivalents reacting $Q = nF$. Therefore, the electrical work obtained from any reaction supplying nF coulombs of electricity at a potential E is,

Net electrical work = nFE

But any work performed by a cell can be accomplished only at the expense of a decrease in free energy occurring within the cell. Further, when the electrical work is a maximum, as when the cell operates reversibly, the decrease in free energy, $-\Delta G$, must equal the electrical work done. We obtain therefore, $\Delta G = -nFE$ from which we see that the reversible EMF of any cell is determined by the free energy change of the reaction going on in a cell. This equation is the bridge between thermodynamics and electrochemistry. Through it the evaluation of various thermodynamic functions of reactions becomes possible from EMF measurements.

Procedure: Standardize the potentiometer. Place the H-shaped cell with Zn (Hg) electrode in one limb, the Ag/AgCl electrode in the other and the $ZnCl_2$ + HCl solution as the common electrolyte in the thermostat. Find out the cell EMF corresponding to the constant temperature of the water bath. Vary the temperature and find out the EMF of the cell at different constant temperatures.

Treatment of Results:

Record the emf measured at different temperatures in the following table.

Temperature (°C)	EMF in volts	ΔH in kJ	ΔS in J	ΔG in kJ

1. Plot EMF against temperature and obtain the slope; slope = $(\delta E/\delta T)_P$ = ... volt/deg.
2. Calculate ΔH and ΔS for the reaction

   $$Zn + 2AgCl \Rightarrow ZnCl_2 + 2Ag$$

   from the following relationships.

   $\Delta S = nF\,(\delta E/\delta T)_P,\qquad \Delta G = -\,nFE$

   $$\Delta H = -\,nF\,[E - T\,(\delta E/\delta T)_P]$$

   where n = 2; F = 96500 coulombs.

Results:

1. Report the temperature coefficient of EMF.
2. Report the values of the thermodynamic functions, ΔG, ΔH and ΔS.

Note:

The change in free energy, ΔG for any process is related to the change in heat content, ΔH by Gibbs-Helmholtz equation

$$\Delta G - \Delta H = T\,\{d(\Delta E)/dT\}_P \qquad (1)$$

If the equation, $\Delta G = -\,nFE$ (2)

If differentiated with respect to temperature then

$$\{d(\Delta E)/dT\} = -\,nF\left(\frac{\partial E}{\partial T}\right)_P \qquad (3)$$

and on substituting equations (2) and (3) into equation (1) we find

$$-\,nFE - \Delta H = -\,nFT\left(\frac{\partial E}{\partial T}\right)_P$$

$$\Delta H = -\, nFE + nFT\left(\frac{\partial E}{\partial T}\right)_P$$

$$= nF\left(\frac{\partial E}{\partial T}\right)_P - E]J \tag{4}$$

Equation (4) permits the calculation of the heat of a reaction from the measured EMF and temperature coefficient of EMF of the reaction, or $(\delta E/\delta T)_P$ is obtained from ΔH and E. In this equation ΔH is in joules if E and F area in volts and coulombs respectively, while ΔH must be substituted in joules to obtain E in volts or $(\delta E/\delta T)_P$ in volts per degree.

Equation (1) indicates that in general ΔF will differ from ΔH by T $(dF/dT)_P$ or in view of equation (3)

$$\Delta G - \Delta H = -\, nFT\left(\frac{\partial E}{\partial T}\right)_P \tag{5}$$

Therefore, for an electrochemical reaction ΔG can equal ΔH only when $(\delta E/\delta T)_P = 0$. Further,

$$\Delta G = \Delta H - T\Delta S \tag{6}$$

Comparison of equations (5) and (6) yields

$$-\, T\Delta S = -\, nFT\left(\frac{\partial E}{\partial T}\right)_P$$

and

$$\Delta S = nF\left(\frac{\partial E}{\partial T}\right)_P$$

EXPERIMENT 64
Formal Potential of a Redox Couple

Aim: To determine the formal potential of a redox couple, $Fe(CN)_6^{3-}$/ $Fe(CN)_6^{4-}$ in different media.

Apparatus Required: Potentiometric assembly, platinum electrodes, connecting wires, KCl salt bridge, 100 cm^3 beakers, standard flasks.

Chemicals Required: 0.1M and 0.01M solutions of $K_4[Fe(CN)_6]$ and $K_3[Fe(CN)_6]$ in 1M H_2SO_4 and also in 1M HCl.

Principle: The redox reaction involving ferrocyanide and ferricyanide ions can be represented as

$$[Fe(CN)_6]^{3-} + e^- \rightleftharpoons [Fe(CN)_6]^{4-} \quad : \quad E^0 = +0.356 \text{ V}$$

The potential, E of the cell for this reaction can be expressed using Nernst equation as,

$$E = E^0 - \{RT/nF\} \log a_{Fe}{}^{2+}/a_{Fe}{}^{3+}$$

or

$$E = E^0 - 0.059 \log (a_{Fe}{}^{2+}/a_{Fe}{}^{3+}), \qquad (\therefore n = 1)$$

$$= E^0 - 0.059 \log (\gamma_{Fe}{}^{2+}/\gamma_{Fe}{}^{3+}) - 0.059 \log \{[Fe^{2+}]/[Fe^{3+}]\}$$

where γ are the respective activity coefficients.

or

$$= E_f^0 - 0.059 \log\{[Fe^{2+}]/[Fe^{3+}]\},$$

where E_f^0 is the potential at unit concentrations of the uncomplexed ions and is called the formal potential of the redox couple. Consequently the cell potential, when the $[Fe^{2+}] = [Fe^{3+}]$ corresponds to formal potential of the redox couple.

The required cell is: $^+Pt \mid Fe(CN)_6^{4-}, Fe(CN)_6^{3-} \mid\mid \text{calomel}^-$

Procedure: Standardise the potentiometer. Take 20 cm^3 of each of the 0.1M $K_4[Fe(CN)_6]$ and $K_3[Fe(CN)_6]$ solutions in 1M H_2SO_4 and mix in a beaker. Place the platinum electrode in this mixture. Connect this half-cell to a saturated calomel electrode by means of a KCl salt bridge. The platinum electrode is connected to the positive terminal of the potentiometer and calomel electrode to the negative end of the potentiometer. Measure the EMF of the cell. Carry out the above measurements with the solutions of lower concentration. Repeat the experiment in 1M HCl instead of H_2SO_4.

Treatment of Data:

$[K_4[Fe(CN)_6]]$ M	$[K_3[Fe(CN)_6]]$ M	Emf measured in volt	
		1M H_2SO_4	1M HCl

For the cell employed,

$$E_{cell} = E_{redox} + E_{calomel}$$

$E_{calomel}$ is the oxidation potential of calomel electrode.

$$\therefore \quad E_{cell} = E_{redox} + 0.242 \text{ volt}$$

$$\therefore \quad E_{redox} = (E_{cell} - 0.242) \text{ volt} \qquad (1)$$

Since the $[Fe(CN)_6]^{4-} = [Fe(CN)_6]^{3-}$, $E_{redox} = E_f^0$ (as shown in the principle section above).

From the E_{cell} measured, using equation (1) calculate the E_{redox} and hence the formal potential, in the different media employed.

Results:

The formal potential of $[Fe(CN)_6]^{4-}$ / $[Fe(CN)_6]^{3-}$ redox, Couple in

(i) 0.1 M of 1M H_2SO_4 = ...V

(ii) 0.01 M of 1M H_2SO_4 = ...V

(iii) 0.1 M of 1M HCl = ...V

(iv) 0.01 M of 1M HCl = ...V

EXPERIMENT 65
pH of a Buffer Solution

Aim: To determine the pH of a series of buffer solutions.

Apparatus Required: Potentiometer assembly, standard cell, calomel electrode, 100 cm^3 beakers, standard flasks, plastic wires, storage battery.

Chemicals Required: 0.2 N acetic acid, 0.2 N sodium acetate and quinhydrone.

Principle: To measure the $[H^+]$ of an acid-base buffer solution, we need to employ an electrode that is reversible with respect to H^+. Quinhydrone, which is an equimolar mixture of quinone and hydroquinone serves this purpose. In the saturated solution, it essentially exists as

The reaction occurring in the cell is

$$\underset{\text{Quinone}}{C_6H_4(OH)_2} \longrightarrow \underset{\text{Hydroquinone}}{C_6H_4O_2} + 2\,H^+ + 2\,e^-$$

$$E = E^0 + (0.0591/2)\log[C_6H_4O_2]\,[H^+]^2/[C_6H_4\,(OH)_2]$$

Since in a saturated solution $[C_6H_4O_2] = [C_6H_4(OH)_2]$, $E^o = 0.6996$ V.

$$\therefore \quad E = 0.6996 + (0.0591/2)\log[H^+]$$

Hence this can be used to determine the pH of solutions

$$-E = -0.6996 - \frac{0.591}{2}\log[H^+]$$

$$\therefore \quad pH = (-E + 0.6996)/\frac{0.0591}{2}$$

where $E = E_{observed} - E_{calomel}$ $E_{calomel} = -0.2415$ V = Oxidation potential of calomel electrode.

Procedure: Standardize the given potentiometer. Prepare the following solutions.

1. 10 cm^3 of 0.2 N CH_3COOH + 20 cm^3 of 0.2 N CH_3COONa and dilute to 100 cm^3.
2. 10 cm^3 of 0.2 N CH_3COOH + 10 cm^3 of 0.2 N CH_3COONa and dilute to 100 cm^3.

3. 20 cm^3 of 0.2 N CH_3COOH + 10 cm^3 of 0.2 N CH_3COONa.

Take the buffer solutions in a beaker and add quinhydrone (half a spatula) and insert a platinum electrode into the solution. The Platinum electrode is connected to the positive end of the potentiometer and this was combined with a calomel electrode through a salt bridge. The calomel electrode is connected to the negative end of the potentiometer. Measure the EMF. Repeat the measurements for all solutions.

Treatment of Results:

pH $= (-E_{observed} + E^{o}_{QH} + E_{calomel}) / 0.0592/2$; $E^{o}_{QH} = 0.6996$ volts;

$E_{calomel} = 0.2415$ volts; $E_{observed}$ = observed EMF

Volume of 0.2 N acetic acid (cm^3)	Volume of 0.2 N sodium acetate (cm^3)	$E_{observed}$ volt	pH (cal)	pH = pK_a + log [(salt/acid)]

pK_a of acetic acid = ...

Result:

Report the pH of the given buffer solutions. Compare the values obtained with the calculated values.

Note:

1. This quinhydrone electrode cannot be used beyond pH=8 or 9 for in alkaline solution

$$C_6H_4(OH)_2 + OH^- \longrightarrow C_6H_4O_2^{2-} + 2H_2O$$

Hence $[C_6H_4(OH)_2] \neq [C_6H_4O_2]$. Therefore this cannot serve as an indicator electrode.

2. It also cannot be used when the electrolyte concentration is too high for the solubilities if Q and HQ are affected.
3. Boron and its compounds should not be present for they form complexes with HQ.
4. For a weak acid HA dissociating as, HA $\rightleftharpoons$ $H^+ + A^-$ the dissociation constant K_a is given by

$$K_a = [H^+]\,[A^-]/[HA]$$

$$\therefore\ [H^+] = K_a\,[HA]/[A^-]$$

$$\therefore\ pH = pK_a + \log\,[A^-]/[HA]$$

EXPERIMENT 66

Solubility Product and the Instability Constant

Aim: To determine the solubility product of AgCl and to determine the instability constant of Ag $(NH_3)_2^+$ complex.

Apparatus Required: Potentiometer assembly, silver electrodes, beakers, standard flasks.

Chemicals Required: 0.01M $AgNO_3$ solution, NH_3 (aq) solution.

Principle: The stability or instability constant of a relatively stable complex ion can be determined by means of EMF measurements. For the general case

$$M_q A_r \longrightarrow q\, M^+ + r\, A^-$$

the instability constant K_I can be represented by

$$K_i = a^q_{M+}\, a^r_{A^-} / {}^aM_qA_r$$

Therefore, $$a^q_{M+} = K_i a_{M_q} A_r a^r_{A^-}$$

If an electrode of the metal M is inserted in the solution of the complex ion, the reversible potential should be given by

$$E = E^o_{M,\,M+} - (RT/qzF)\ ln\ K_i - (RT/qzF)\ ln\ a_{M_q} A_r / a^r_{A^-}$$

It is therefore possible to determine the instability constant of the silver ammonia complex from the measurement of the EMF of a suitable cell.

The solubility product can be determined from the measurement of the EMF of a suitable electrochemical cell from which the respective activities of the ions in the saturated solution of the sparingly soluble salt can be determined and used in the calculation of the solubility product of the salt.

Procedure:

I. Determination of solubility product K_{sp} of AgCl:

Set up a cell as follows:

$$Ag/AgNO_3(0.1M) // KC1(0.1M),\ AgCl\ (satd)/Ag$$

Measure the EMF. $E = RT \text{ in } (a_{Ag^+})_1 / (a_{Ag^+})_2$

where $(a_{Ag^+})_1$ is the activity of 0.1M Ag^+ solution and $(a_{Ag^+})_2$ is the activity of Ag^+ present in the saturated solution of AgCl. $(a_{Ag^+})_1$ is known from the tables. Hence $(a_{Ag^+})_2$ can be calculated. Activity of Cl^- (a_{Cl^-}) is also known from the tables at the appropriate concentration (in this case at 0.01M concentration).

Therefore $K_{sp} = a_{Ag+} \times X\ a_{Cl^-}$ can be calculated.

II. Determination of instability constant of $Ag(NH_3)_2^+$ complex:

Set up the cell Ag/$AgNO_3$ (0.01M)/ /$AgNO_3$ (0.025 M) in 1M NH_3/Ag. Measure the EMF using a standardized potentiometer. The measured EMF is

$$E = 0.0591 \log (a_{Ag^+})_1 / (a_{Ag^+})_2$$

where $(a_{Ag^+})_2$ is the activity of Ag^+ ions in 0.01M $AgNO_3$ solution and $(a_{Ag^+})_2$ is the activity of Ag^+ ions in 0.025 M $AgNO_3$ solution in 1M ammonia.

Since $K_i = \dfrac{\left[a_{Ag+} \times a^2_{NH_3}\right]^2}{\left[a_{Ag(NH_3)_2}\right]}$, Instability constant can be calculated.

Treatment of Data:

Titration of 10 cm^3 of KCl versus $AgNO_3$

Volume of $AgNO_3$ (cm^3)								
EMF (V)								

Volume of KCl = 10 cm^3

Volume of $AgNO_3$ required (from graph) = ...(a) cm^3

Strength of $AgNO_3$ = 0.1M

Therefore Strength of KCl = 0.1 × a

Determination of the instability constant of $Ag(NH_3)_2^+$

EMF of the cell measured = ... volt

$$E = 0.0591 \log \left(a_{Ag^+}\right)_1 / \left(a_{Ag^+}\right)_2$$

Therefore $$a_{[Ag(NH_3)_2]^+} = c_{[Ag(NH_3)_2]^+} \gamma_{[Ag(NH_3)_2]^+}$$

$$K_i = \frac{a_{Ag+} \times \left(a_{NH_3}\right)^2}{a_{[Ag(NH_3)_2]^+}}$$

$$a_{[Ag(NH_3)_2]^+} = c_{[Ag(NH_3)_2]^+} \times \gamma_{[Ag(NH_3)_2]^+}$$

Results:

1. Report the solubility product of AgC1
2. Instability constant of $Ag(NH_3)_2^+$ complex.

EXPERIMENT 67
Activity of Hydrogen Ions Using the Hydrogen Electrode

Aim: To determine the activity of hydrogen ions in acid medium using the hydrogen electrode hence to determine the ionic product of water and the hydrolysis constant of sodium acetate.

Apparatus Required: Potentiometer assembly, Hydrogen gas electrode, beakers and standard flasks.

Chemicals Required: Hydrogen gas, sodium hydroxide, pure distilled water.

Principle: The EMF of a cell free from liquid junction potential consisting of a hydrogen electrode and a reference electrode is given by,

$$E = E_{ref} - (RT/F)\ln a_{H+}$$

or

$$E = E_{ref} + [(2.303\ RT)/F]\ pH$$

Hence

$$pH = (E - E_{ref})/0.0591$$

Since pH = – log a_{H+}, one can calculate the activity coefficient of hydrogen ions.

The EMF of the cell,

$$H_2(gas)/KOH\ (0.01N)/\ /HCl\ (0.01N)/H_2\ (1\ atm)$$

for which it is supposed that the liquid junction potential has been completely eliminated is given by

$$E = (RT/F)\ \ln\ (a'_{H+}/a''_{H+})$$

where a'_{H+} and a''_{H+} represent the hydrogen ion activities in the right hand and left hand solutions respectively. i.e., in the 0.01 N HCl and 0.01 N KOH solutions. If a''_{OH} is the hydroxyl ion activity in the latter solution then,

$$a''_{H^+} \cdot a''_{OH^-} = K_w$$

since $E = (RT/F) \ln [(a'_{H^+} \times a''_{OH^-})/K_w]$, K_w can be calculated.

Sodium acetate is hydrolysed as follows:

$$CH_3COO^- + H_2O \rightleftharpoons CH_3COOH + OH^-$$

Applying the law of mass action,

$$K_h = a_{HA}A_{OH^-}/a_{A^-}$$

In terms of the degree of hydrolysis 'α' this can be written as $K_h = \alpha^2 c/(1 - \alpha)$, hence K_h can also be calculated.

Procedure:

Part 1. Determination of activity coefficient of hydrogen ions:

Prepare 0.1N, 0.01N and 0.001N solutions of HCl.

Set up the following cell

$$Pt, H_2 (g)/HCl \text{ solution}/ /\text{Calomel electrode}$$

and measure the EMF for different concentrations of HCl solution.

$$pH = (E - 0.2415)/0.0591$$

therefore, pH can be calculated.

Since $pH = -\log a_{H^+}$, the activity coefficients of hydrogen ions can be calculated. Since the activity coeffcient is given by $\Upsilon_{H^+} = a_{H^+}/c$, where c is concentration, Υ_{H^+} at various values of c can also be calculated.

Part 2. Determination of ionic product of water:

Prepare an exactly 0.1N solution of NaOH.

Measure the EMF of the cell,

$$Pt, H_2(g)/NaOH \text{ (0.1N solution)} / / \text{calomel electrode}$$

since we already know the half cell potential for the half cell Pt, $H_{2(g)}$/HCl (0.1N), we can obtain the EMF of the following cell

$$Pt, H_{2(g)}/HCl/0.1N\ HCl/ /0.1N\ NaOH/H_2(g).\ Pt$$

Using the equation

$$E = (RT/nF) \ln (a_{H^+} \cdot a_{OH^-})/K_w$$

The ionic product of water can be calculated.

Part 3. Determination of the hydrolysis constant of sodium acetate:

Prepare 0.1 N and 0.01 N solutions of sodium acetate.

Measure the EMF of the following cell

Pt, H_2 (g)/NaOAc solution/ /calomel electrode

$$pOH = 14 - [(E_{obs} - 0.2415)/0.0591]$$

$$pOH = -\log a_{OH}$$

Therefore, degree of hydrolysis, $\alpha = a_{OH}/c$. and hydrolysis constant $K_h = \alpha^2 c/(1 - \alpha)$, and K_h can be calculated.

Treatment of Data:

I. Determination of activity coefficient of H^+ ions:

Concentration of HCl	EMF volt	pH	a_{H^+}	Υ_{H^+}
0.1 N 0.01 N 0.001 N				

Cell Pt, H_2(g)/HCl solution/ /calomel electrode

$pH = (E - 0.2415)/0.0591$ $pH = -\log a_{H+}$ $\Upsilon_{H^+} = a_{H^+}/c$

II. Determination of ionic product of water

Cell Pt, H_2(g)/NaOH solution/ /calomel electrode

$$(E_{NaOH} - E_{HCl})/0.0591 = \log (a_{H^+} \cdot a_{OH^-})/K_w$$

III. Determination of hydrolysis constant of sodium acetate

Cell Pt, H_2(g)/Sodium acetate solution/ /calomel electrode

Sodium acetate	EMF volt	pH	pOH	a_{OH-}	$a = a_{OH-}/c$	K_h
0.1 N 0.01 N						

Result:

1. The activity coefficient of H^+ ions at different concentrations.

[HCl] N	0.1	0.01	0.001
a_{H+}			

1. Ionic product of water = ...
2. The value of the hydrolysis constant of sodium acetate = ...

Note:

Construction and Working of Hydrogen electrode:

The hydrogen gas electrode consists of a small platinum sheet or wire coated with finely divided platinum black by electrolysis of chloroplatinic acid containing a small portion of lead acetate. Thc Platinum foil or wire attached to a suitable connecting wire is inserted in the experimental solution through which a stream of hydrogen gas is passed at atmospheric pressure. The position of the electrode in the solution is arranged so that it is partly in the atmosphere of the hydrogen gas. A simple and convenient type of hydrogen electrode is that, usually associated with the name of Hildebrand' a rectangular sheet of platinum of about 1 to 3 cm^2 exposed area, which is subsequently platinized is welded to a short length platinum wire sealed into a glass tube containing mercury. This tube is sealed into another tube closed at the top, but widening out into a bell shape in the region surrounding the platinum sheet; a side connection is provided for the inlet of hydrogen gas. A number of holes or slits are made in the bell shaped portion of the tube at a level midway up the platinum so that when the electrode is inserted in a solution and hydrogen passed in through the side tube the platinum sheet is half immersed in liquid and half is surrounded by gas. This arrangement permits the rapid attainment of equilibrium between the electrode material, the hydrogen gas and solution. The time taken to reach this state of equilibrium depends, among other factors, on the nature of the solution, the thickness of the deposit and on the previous history of the electrode. As a general rule, an electrode that is functioning in a satisfactory manner will give a steady potential within five or ten minutes commencing the passage of hydrogen. The hydrogen gas should be purified by bubbling it through alkaline permanganate and alkaline pyrogallol solutions to remove oxygen and other impurities which may influence the functioning of the hydrogen electrode.

Whatever form of electrode vessels is employed, the fundamental principle of operation is always the same. The hydrogen gas is adsorbed by the finely divided platinum and this permits the rapid establishment of equilibrium between molecular hydrogen on the one hand, and hydrogen ions in solution and electrons, on the other hand, thus,

$$1/2\ H_2\ (g) \rightleftharpoons 1/2\ H_2(Pt) + H_2O \rightleftharpoons H_3O^+ + e^-$$

This equilibrium can be attained rapidly from either direction and so the electrode behaves as one that is reversible with respect to hydrogen ions.

EXPERIMENT 68
Hydrolysis Constant and the Degree of Hydrolysis

Aim: To determine the degree of hydrolysis and the hydrolysis constant of anilinium hydrochloride by potentiometry.

Apparatus Required: Potentiometry assembly, Platinum electrode and calomel electrode.

Chemicals Required: Anilinium hydrochloride, quinhydrone.

Principle: Anilinium hydrochloride, $C_6H_5NH_3^+Cl^-$ when dissolved in water, ionizes to form $C_6H_5NH_3^+$ and Cl^- ions, and the cation establishes the following hydrolytic equilibrium.

$$C_6H_5NH_3^+ + H_2O \rightleftharpoons C_6H_5NH_2 + H_3O^+$$

The equilibrium constant for this hydrolytic process is called the hydrolysis constant for the salt and is given by, $K_h = (a_{H+} \times a_B)/a_{BH+}$ where a_{H+} is the activity of the free acid (H_3O^+); a_B is the activity of the free base ($C_6H_5NH_2$), and $a_{BH}+$ is the activity of the unhydrolysed salt ($C_6H_5NH_3^+Cl^-$). However, for dilute solutions, we may replace activities by concentration terms;

Hence,

$$K_h = [H^+]\,[B]/[BH^+] \tag{1}$$

Hydrolysis constant can also be related to the dissociation constant, K_b of the base through the ionic product of water, K_w as

$$K_h = K_w/K_b \tag{2}$$

If c equivalents of the salt is dissolved in a litre of water, cα equivalents of each free base and free acid will be formed due to hydrolysis (α is the degree of hydrolysis).

Thus, the pH of the solution may be related to the degree of hydrolysis as,

$$pH = -\log[H^+] = -\log(c\alpha)$$

Hence, by measuring the pH of the solution, $c\alpha$ can be calculated from which the degree of dissociation, α can be obtained at a given concentration.

Also, expressing K_h in terms α, using $K_h = c\alpha^2/(1-\alpha)$, the hydrolysis constant can be calculated. Substituting for K_h in equation (2) and taking $K_w = 1.0 \times 10^{-14}$ at 25° C, the dissociation constant of the base, K_b can be evaluated.

Procedure: Standardise the potentiometer. Prepare an M/10 aniline hydrochloride solution by dissolving appropriate quantity of the substance in distilled water (100 cm³). From this stock solution, dilute appropriately and get M/20, M/50 and M/100 solutions. Then construct the following cell:

$^{+}$Pt/0.1M anilinehydrochloride quinydrone / / calomel^{-}

Determine the potential of the cell. Repeat the experiment with each of the other solutions.

Treatment of Data:

1. pH is given by pH = $(-E_{obs} + E_{QH} + E_{cal}) / 0.0591$ where $E_{QH} = 0.6996$ V and $E_{cal} = -0.242$ V (Oxidation Potential). From this relation, pH of the solution can be calculated.
2. As pH = $-\log[H^+] = -\log(c\alpha)$, pH = $-\log c - \log\alpha$, the degree of hydrolysis α can be calculated at every given concentraion.
3. From α, calculate the hydrolysis constant using the relation, $K_h = c\alpha^2/(1-\alpha)$
4. The Dissociation constant, K_b can be calculated from the relation $K_b = K_w/K_h$

Tabulation:

$C_6H_5NH_3^+Cl$	E_{obs} (V)	pH	α	K_h	K_b
N/10					
N/20					
N/50					
N/100					

Mean K_h = – ..., Mean K_b = ...

Result:

Report the values of hydrolysis constant (K_h) and dissociation constant (K_b).

EXPERIMENT 69
Titration of a Strong/Weak Acid with a Strong Base

Aim: To determine the strength of the given acid by potentiometric titration.

Apparatus Required: Potentiometry assembly, platinum electrode, calomel electrode and connecting wires.

Chemicals Required: A decinormal solution of HCl, standardized NaOH solution (about 0.1N), quinhydrone.

Principle: When an electrode is dipped in an electrolyte solution, its potential depends on the concentration of the ions to which the electrode is reversible. In an acid-base titration, quinhydrone (QH) which is an equimolar mixture of quinone and hydroquinone is employed as H^+ reversible system, and the following cell is constructed:

Pt/acid + (QH) / / calomel

The reduction potential of quinhydrone and the emf of the cell will depend on the pH of the solution. On adding small aliquots of a base the emf of the cell drops gradually; but, near the equivalence point a large decrease in emf occurs. On crossing the equivalence point, again emf changes in small decrements.

Procedure: Pipette out 20 cm^3 of the given acid into a clean beaker. Add a pinch of quinhydrone and place a platinum electrode in the solution. Connect it to a calomel electrode through a salt bridge. Run 1 cm^3 aliquots of the standard NaOH from the burette and determine the emf after every addition. At one point when the emf measurement becomes difficult, reverse the poles and continue measurement for a few more aliquots addition. The equivalence point lies in the range where the electrodes are reversed.

This is the pilot titration. Then repeat the experiment more carefully by adding $0.1 cm^3$ at a time in the range of the equivalence point.

Treatment of Data:

Strength of NaOH = ...N

Volume of HCl = 20 cm^3

1. Pilot run: (1cm^3 addition)

Volume of NaOH (cm^3)									
emf (V)									

2. Actual run: (0.1cm^3 addition)

Vol. of NaOH cm^3	emf (V)	ΔE (V)	ΔV (cm^3)	ΔE/ΔV (V/cm^3)

Draw a graph of emf Vs volume of NaOH added and dE/dV Vs. volume of NaOH added.

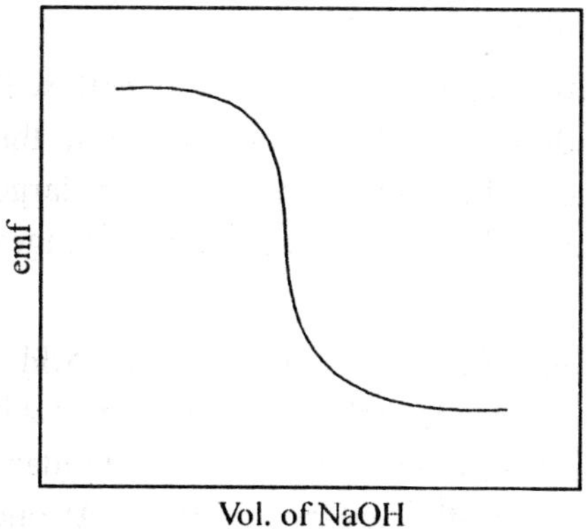

Fig. 36

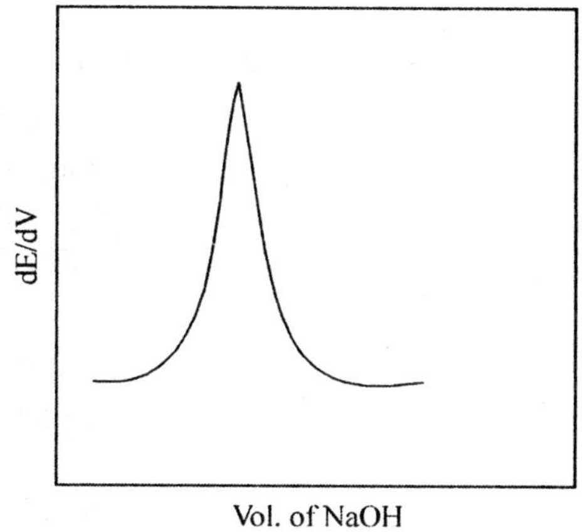

Fig. 37

Results:

(i) The volume of base required to neutralize the acid = ... cm^3.

(ii) The strength of HCl = ... N.

EXPERIMENT 70

Titration of a Strong and Weak Acid Mixture with a Strong Base

Aim: To determine the concentrations of a strong acid and a weak acid present in a mixture by potentiometric titration.

Apparatus Required: Potentiometer assembly , platinum electrode, calomel electrode and connecting wires.

Chemicals Required: An acid mixture consisting of approximately 0.1 N HCl and 0.1 N acetic acid; standardized NaOH solution (about 0.1 N); quinhydrone.

Principle: The fundamental principle is the same as given in Experiment 69. The cell is to be constructed as shown below:

Pt/acid mixture + QH/ /calomel

On adding small aliquots of an alkali, first the strong acid reacts and the emf of the cell gradually decreases, and when strong acid is fully neutralized the emf decreases significantly. Further, weak acid reacts with the alkali and hence, again *emf* decreases gradually. As the weak acid is also fully neutralized, further addition of alkali renders the solution basic and the poles reverse, indicating completion of neutralization reaction.

Procedure: Pipette out 20 cm^3 of the given acid mixture into a clean beaker. Add a pinch of quinhydrone and place a platinum electrode in the solution. Connect it to a calomel electrode through a salt bridge. Run 1 cm^3 aliquots of the standard NaOH from the burette and determine the emf after every addition. The emf initially decreases gradually, dips suddenly and continue to decrease gradually again. Note the Volume (V_A) of alkali added when the emf dips significantly. Continuing addition of alkali aliquots further, leads to a pole reversal. Note the volume (V_B) of alkali added when the pole is reversed. This is the pilot run.

Then wash the beaker and repeat the experiment by taking 20 cm^3 of the given acid mixture and adding 0.1 cm^3 of NaOH first in the neighbourhood of V_A and then near V_B.

Treatment of Data:

Strengh of NaOH = ... N.
Volume of acid mixture = 20 cm^3.
1. Pilot run: (1 cm^3 addition)

Vol. of NaOH (cm^3)	1	2	3	4	5	6	–	–	–
EMF (volt)									

2. Actual run: (0.1 cm^3 addition)

Vol. of NaOH (V) cm^3	EMF, E	dE	dV cm^3	dE/dV	V/cm^3

Draw a graph of dE/dV values against volume of alkali added (Fig. 38). The volume (a) corresponding to the first peak gives the volume of NaOH at the equivalence point of HCl neutralization and the volume, b corresponds total NaOH required for complete neutralization. Hence (b – a) corresponds to volume of NaOH for acetic acid.

From the respective volume of NaOH consumed, calculate the concentration of HCl and acetic acid.

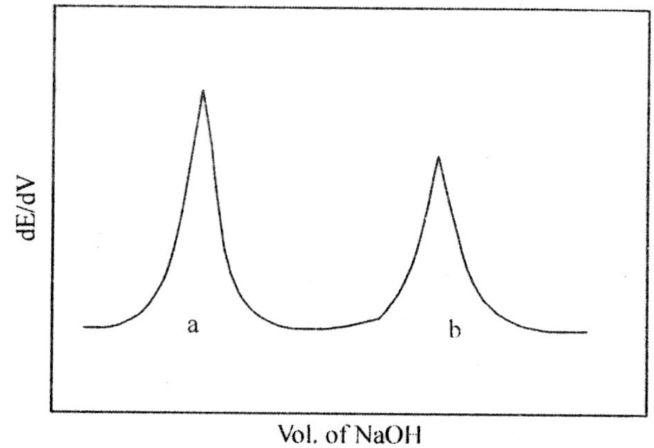

Fig. 38

Results:

(i) The volume of NaOH required for HCl = ...(a).

The volume of NaOH required for CH_3COOH = ... (b – a).

(ii) The Strength of HCl = ... N.

The strength of CH_3COOH = ... N.

EXPERIMENT 71
Dissociation of a Weak Acid by Potentiometric Titration

Aim: To determine the dissociation constant of a weak acid by potentiometry.

Apparatus Required: Precision pH–meter, glass elcctrode and calomel electrode, standard buffer solutions, pipettes, graduated flasks, beakers.

Chemicals Required: Stock solutions of acetic acid (0.1 M), NaOH (0.1 M) and NaCl (0.2 M).

Principle: The weak acid exists in the following dissociation equilibrium in water

$$HA + H_2O \longleftrightarrow H_3O^+ + A^-$$

For which the dissociation constant is given by

$$K_a = [a_{H+} \cdot a_{A^-}]/a_{HA}$$

where 'a' refers to the corresponding activities.

The value of the acid dissociation constant decides whether the given acid is a strong acid. The acidity is measured in the pH scale defined and $pH = -\log a_{H+}$ and hence one can also define a $pK_a = -\log K_a$

Therefore one rewrites the dissociation constant of a weak acid as follows

$$pK_a = pH + \log [HA]/[A^-]$$

and pK_a is a measure of the strength of acid, and weak acids have pK_a values > 2.

Consider the neutralization of a weak acid HA by a strong base like NaOH and the effective neutralization reaction can be written as

$$HA + OH^- \rightleftharpoons H_2O + A^-$$

and therefore one can write the following equation for the concentrations of HA and A^- at any point of the neutralization

$$[HA] = [HA]_{initial} - (NaOH)_{added}$$

$$[A^-] = (NaOH)_{added}$$

Prepare the following set of solutions

Sample/solution	1	2	3	4	5	6	7	8	9
Acetic acid	20	20	20	20	20	20	20	20	20
NaOH	2	4	6	8	10	12	14	16	18
NaCl	12	11.5	11	10.5	10	9.5	9	8.5	8
H_2O	16	14.5	13	11.5	10	8.5	7	5.5	4

Now set up the pH meter and connect the glass electrode and the calomel electrode. Use a standard buffer and calibrate the pH meter. After that wash the electrode and dry. Now record the pH of each of the solution giving sufficient time for the electrodes to respond to the environment.

Treatment of Data:

1. Calculate the analytical concentration of acetic acid and also that of NaOH in each of the solution from the values of the stock solutions.
2. Calculate the equilibrium concentrations of the acid and the base $[CH_3COOH]$, $[CH_3COO^-]$ using the following equations

$$[CH_3COOH] = (CH_3COOH)_{initial} - (NaOH)$$

$$[CH_3COO^-] = (NaOH)_{added}$$

3. Calculate the pK_a of the given acid using the equation

$$pK_a = pH + \log [CH_3COOH]/[CH_3COO^-]$$

for each of the solutions.

4. Calculate the value of K_a from the values of pK_a obtained. Calculate the mean value of K_a using the equation

$$<K_a> = (\Sigma\ 10^{pKa})/N$$

where N is the number of pK_a values used for the calculation of the mean value.

5. Plot the neutralization diagram (pH versus volume of NaOH added) and read the value of pH corresponding to the point where the neutralization is 50 %.

Result:

Report the value of the K_a of the acetic acid and also calculate the relative error of your measurement from the tabulated value of K_a.

EXPERIMENT 72
Redox Titration

Aim: To determine the concentration of a reductant or an oxidant by a potentiometric redox titration.

Apparatus Required: Potentiometer, calomel electrode, beakers, standard flask, and platinum electrode.

Chemicals Required: 0.1 N ferrous ammonium sulphate and 0.1 N ceric sulphate — both prepared in 2 N sulphuric acid.

Principle: Consider an electrochemical cell of the type

$$Hg \mid Hg_2Cl_2 \mid \text{Satd. KCl} \mid\mid Fe^{2+}, Fe^{3+} \mid Pt$$

Addition of ceric sulphate would alter the relative concentration of Fe^{3+}/Fe^{2+}. Ferric/ferrous half cell potential is given by

$$E = E^0 + 0.0509 \log [Fe^{3+}]/[Fe^{2+}]$$

where $E^0 = 0.77$ volt

As ferric concentration increases, the ferrous concentration decreases and the half cell potential as well as the cell potential increases. Near the equivalence point, the change in potential will be higher as the ferric/ferrous ratio changes considerably. However, after the equivalence point, only ferric ions are present and since no ferrous ions are present in solution, one has another redox couple, namely, ceric/cerous system whose half cell potential is given by

$$E = E^0 + 0.509 \log [Ce^{4+}]/[Ce^{3+}]$$

where $E^0 = 1.61$ volt.

Hence the potential after the equivalence point is governed by the ceric/cerous ratio available in the solution. The sudden change in potential from one system to another can be used to determine the equivalence point.

Procedure: Set up the following cell using a platinum wire.

$$Hg \mid Hg_2Cl_2 \mid KCl\ (Satd.) \mid\mid Fe^{2+}, Fe^{3+} \mid Pt$$

using a known volume of ferrous ammonium sulphate solution. Connect the platinum electrode to the positive end and the calomel electrode to the negative end of the potentiometer. Record the emf of the cell as a function of the addition of ceric sulphate solution taken in the burette. Add every time 1 to 2 cm^3 portions and stir the solution thoroughly before recording the EMF.

Treatment of Results:

Volume of ferrous ammonium solution = . . . cm^3.
Strength of ceric sulphate solution = . . . N.
Volume of ceric sulphate at the equivalence point = . . . cm^3.
Strength of the given ferrous ammonium sulphate solution = . . . N.

Vol. of ceric-sulphate (cm^3)									
emf (volt)									
$\Delta E/\Delta V$									

1. Plot emf versus volume of ceric sulphate solution added. Note the base potential as well as the value of potential after the equivalence point and check whether they correspond to cell potentials calculated on the basis of ferric/ferrous or ceric/cerous couple.
2. Plot $\Delta E/\Delta V$ versus volume of ceric sulphate added. Identify the equivalence point.

Result:

Report the strength of the given ferrous ammonium sulphate solution.

EXPERIMENT 73
Composition of Zinc Ferrocyanide Complex by Potentiometric Titration

Aim: To determine the composition of zinc ferrocyanide complex potentiometrically.

Apparatus Required: Potentiometer assembly, platinum electrode, calomel electrode, KCl bridge, beakers.

Chemicals Required: 0.1 M solution of zinc sulphate, 0.1 M solution of potassium ferrocyanide and 1 % solution of potassium ferricyanide.

Principle: The oxidation potential of a ferrocyanide – ferricyanide electrode is given by,

$$E = E^0 + 0.591 \log [Fe(CN)_6]^{3-}/[Fe(CN)_6]^{4-}$$

As long as there is excess of ferrocyanide ions in solutions, the potential will be small. On adding zinc sulphate, the concentration of ferrocyanide decreases and the potential increases. At the end point when all the ferrocyanide is quantitatively precipitated as zinc ferrocyanide complex, there is a large increase in potential. Hence the volume of zinc sulphate required to complex with a known concentration of ferrocyanide can be known.

Procedure: Standardize the potentiometer. Pipette out 20 cm^3 of the 0.1 M potassium ferrocyanide solution into a beaker. Add 1 cm^3 of 1 % solution of potassium ferricyanide. Dip a platinum electrode into the solution. Combine this half cell with a saturated calomel electrode through a KCl salt bridge. Connect the platinum electrode to the positive end of the potentiometer and the calomel electrode to the negative end of the potentiometer. Measure the potential of the solution initially as well as after every addition of zinc sulphate.

Treatment of Results:

Titration of $K_4[Fe(CN)_6]$ versus zinc sulphate

Volume of $K_4[Fe(CN)_6]$ solution = 20 cm^3

Volume of zinc sulphate at the end point = cm^3.

20 cm^3 of 0.1 N $K_4[Fe(CN)_6]$ = ... cm^3 of 0.1 M zinc sulphate solution.

Therefore, 2 moles of ferrocyanide = 3 moles of zinc sulphate.

Therefore, the formula of the complex is $K_2Zn_3[Fe(CN)_6]_2$.

The reaction between Zn^{2+} and ferrocyanide ions is

$$3\ Zn^{2+} + 2\ K_4[Fe(CN)_6] \longrightarrow K_2Zn_3\,[Fe(CN)_6]_2 + 6\ K^+$$

Volume of Zinc sulphate solution (cm^3)	EMF (volt)	$\Delta E/\Delta V$
1		
2		
3		
–		
–		
–		

1. Plot $\Delta E/\Delta V$ against volume of zinc sulphate solution added.
2. Determine the end point from the plot. The end point is the volume of $ZnSO_4$ added corresponding to the maximum in the graph.
3. Calculate the ratio in which zinc sulphate and potassium ferrocyanide react to form the zinc ferrocyanide complex.

Result:

The formula of the complex is found to be $K_2Zn_3\,[Fe(CN)_6]_2$.

EXPERIMENT 74

Precipitation Titration of a Mixture of Chloride and Iodide Ions with Silver Nitrate

Aim: To determine the amount of KI and KCl present in a mixture by potentiometric titration.

Apparatus Required: Potentiometer, standard calomel electrode, silver electrode, ammonium nitrate salt bridge.

Chemicals Required: KCl, KI, $AgNO_3$, distilled water.

Principle: The concentration of Cl^- and I^- present in the mixture can be determined by titrating the given mixture against a standard solution of $AgNO_3$. On carrying out the titration, AgI and AgCl are precipitated out as they are sparingly soluble in water. This is a good example for a *precipitation titration.* On adding the titrant, $AgNO_3$ from the burette to the mixture in a beaker , a reversible electrode Ag/AgCl(s), Cl^- is obtained. This is coupled with the calomel reference electrode. Thus the overall cell is Ag/AgCl(s), Cl^-/ / calomel.

The emf of the cell is given by

$$E + E^0_{Ag} - (RT/F) \ln a_{Ag+} - E_{calomel}$$

(i.e.),
$$E = (E^0_{Ag} - E_{calomel}) - (RT/F) \ln a_{Ag+}$$

At 25° C, for dilute solutions we can write the above equation as

$$E = \text{constant} - 0.059 \log [Ag^+]$$

Since the solubility of AgCl and AgI are very less, the changes in emf would also be very small. But at the equivalence point when a slight excess of $AgNO_3$ is added, a sharp change in emf occurs.

Procedure: Standardise the potentiometer. Pipette out 10 cm^3 of KCl and 5 cm^3 of KI (or 10 or 20 cm^3 of the given halide mixture) into a beaker. Dip a silver electrode into this solution. Connect this to the positive end of the potentiometer. Combine this half-cell with a saturated calomel electrode, through an ammonium nitrate salt bridge, which is connected to the negative end of the potentiometer. Titrate the solution against 0.1 N $AgNO_3$ and record the potential after every cm^3 of the addition. Plot E versus volume of $AgNO_3$ as well as dE/dV versus volume of $AgNO_3$ and determine the end points. For more accurate results, the titrant may be added in the order of 0.1 cm^3 aliquots in the neighbourhood of the equivalence point.

Treatment of Data:

Volume of KCl = ...cm^3; Volume of KI = ...cm^3

Concentration of $AgNO_3$ = ...N.

Vol. $AgNO_3$ (cm^3)									
emf (E) (volt)									
$\Delta E/\Delta V$									

Plot the volume of $AgNO_3$ added against $\frac{dE}{dV}$. The graph will have two peaks; from the position of the peaks the exact volume of the titrant required for each halide can be found out.

Result:

Report the concentrations of KCl and KI in the given mixture of the halides.

PART X
Electrode Processes

EXPERIMENT 75
Percentage Purity of Copper Sulphate — Electrogravimetry

Aim: To determine the percentage purity of the given sample of copper sulphate.

Apparatus Required: Electrolysis apparatus with platinum electrodes, volumetric flask – 250 cm^3.

Chemicals Required: Copper sulphate crystals, conc. H_2SO_4, conc. HNO_3, distilled water.

Principle: This experiment is based on the principle of *electrogravimetry*. In this experiment, the amount of copper present in the given copper sulphate sample is deposited on the cathode by electrolyzing the solution of the sample. The deposition occurs quantitatively, as per the Faraday's laws of electrolysis, as

$$Cu^{2+} + 2\ e^- \longrightarrow Cu$$

From the difference in weight of the copper deposited and that initially taken, the purity of the copper sulphate sample can be calculated.

Procedure: Prepare a standard solution of copper sulphate containing 0.5g in 100 cm^3 of water. Take the solution in a 400 cm^3 beaker. Add 2 cm^3 of concentrated sulphuric acid. Clean the Platinum gauze cathode in 1:1 nitric acid, and wash thoroughly with distilled water followed by pure acetone. Dry the electrode at 100 – 110° for 3-4 minutes . Cool it in air and then weigh. This electrode will serve as the cathode. Dip the cathode partially in copper sulphate solution taken in the beaker. Connect the cathode to the negative terminal of the electro-analysis apparatus and the anode to the positive terminal. Adjust the height of the electrodes such that the electrodes extend nearly to the bottom of the beaker and the cathode is ¾ immersed in the solution. Switch on the mains supply and adjust the voltage to 3 volts and maintain a current between

2-4 amperes. Stir the solution by means of magnetic stirrer. Continue the electrolysis for about 15 minutes until the blue colour of the solution has entirely disappeared. Test for completion of deposition of copper by immersing the cathode further into the solution and seeing whether there is any deposition of copper on the fresh surface exposed to the solution. Remove the electrode from the solution carefully. Wash the electrode with distilled water and then with acetone. Dry the electrode at 100-110°C, cool the electrode and then weigh it.

Treatment of Data:

Weight of the cathode before deposition = ...(w_1)

Weight of the cathode after deposition = ... (w_2)

Weight of the copper deposited = ($w_2 - w_1$)

Amount of copper sulphate in the sample = (249.5/63.5) × ($w_2 - w_1$) = y, (say)

Therefore, % purity of the sample = (y/0.5) × 100.

Result:

The % purity of the copper sulphate sample is ....

EXPERIMENT 76
Decomposition Potential

Aim: To measure the decomposition potentials of sulphuric acid, zinc sulphate, sodium sulphate and copper sulphate.

Apparatus Required: Battery, rheostat, voltmeter, ammeter, platinum electrodes, beaker.

Chemicals Required: H_2SO_4, $CuSO_4$, $ZnSO_4$, Na_2SO_4.

Principle: At a reversibe electrode, which is in a state of equilibrium the discharge of ions and their formation take place at the same rate and there is no net flow of current. If however, as the result of the application of an external EMF there is an actual passage of current, the electrode is disturbed from its equilibrium condition and the electrode is said to be polarised. When the applied EMF is just sufficient to disturb the system so that the electrolyte starts decomposing into its ions, it is referred to as the discharge or decomposition potential.

If two small sheets of smooth platinum are immersed in dilute sulphuric acid and a gradually increasing EMF is applied by means of an external battery so as to electrolyze the solution, a galvanometer in the circuit will show that the current increases in the manner represented by the curve shown below (Fig. 40). It appears that an appreciable voltage must be applied before current can flow freely through the cell. The EMF applied at the point D at which, steady electrolysis commences is the decomposition voltage of the particular solution with the given electrode material. It is at this point that in general the steady evolution of hydrogen and oxygen gas bubbles is observed. Attention should be drawn to the fact that the nature of the current-voltage curve is such that D cannot be defined precisely, in any event. It is now believed not to have any exact theoretical significance. The apparent decomposition voltage is nevertheless of interest for it gives an approximate indication of the minimum EMF which must be applied to a particular solution in absence of depolarizers, in order that current of appreciable magnitude shall be able to pass. Hence, it represents the sum of potentials which must be attained by the two electrodes before the rates of the respective ion discharge processes are appreciably greater than the reverse reactions.

Procedure: The circuit is completed as shown in Fig.39.

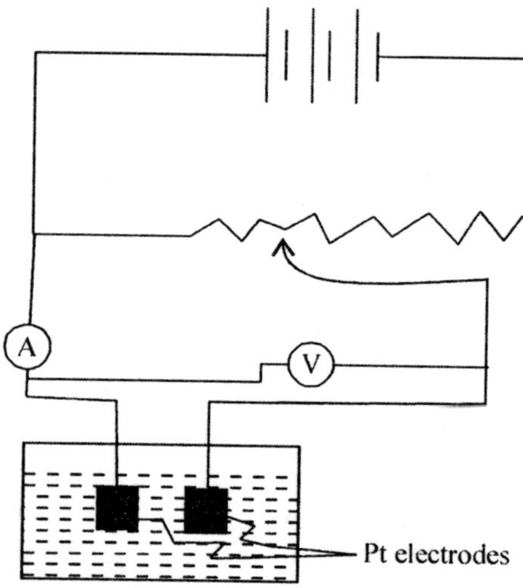

Fig. 39. Circuit for the study of current – voltage curves.

Prepare 0.1 N solutions of sulphuric acid, copper sulphate, zinc sulphate and sodium sulphate. Take sulphuric acid in a beaker and immerse two platinum electrodes in it such that they are parallel to each other, platinised platinum electrodes should be used. Measure the current flowing through the cell at various voltages applied by adjusting the rheostat. Repeat the experiment with the other solutions.

Treatment of Data:

Record the observed data in the following table.

0.1 N H_2SO_4		0.1 N $ZnSO_4$		0.1 N $CuSO_4$		0.1 N Na_2SO_4	
Voltage	Current	Voltage	Current	Voltage	Current	Voltage	Current

Plot the current against voltage (Fig. 40). The point at which there is a sharp increase in current corresponds to the decomposition potential.

The decomposition potentials (from graph) of

Sulphuric acid	= ... volts;	Zinc sulphate	= ... volts
Copper sulphate	= ... volts;	Sodium sulphate	= ... volts

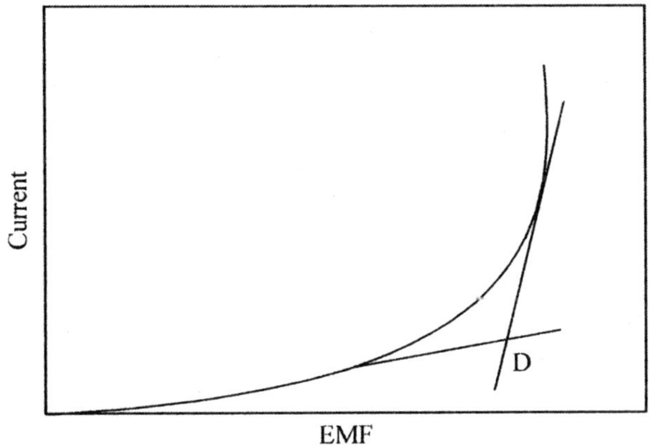

Fig. 40. Typical current-voltage curve.

Result:

Report the decomposition potentials of the various solutions.

Note:

The decomposition potential of acids and bases are in the vicinity of 1.7 volts and hence it can be concluded that the same electrolytic process occurs at the electrodes in each case. The only process which could be common to all aqueous solutions of acids and bases is the discharge of hydrogen ions at the cathode and of hydroxyl ions at the anode to form hydrogen and oxygen respectively. It appears therefore, that although the current is actually carried by all kinds of ions in the electrolyte only certain cations, not necessarily those present in excess are actually discharged at the electrodes. Thus, in sodium hydroxide solution, for example, the current is carried toward the cathode almost exclusively by sodium ions, but it is nevertheless the hydrogen ions that are discharged, at least at a platinum electrode.

The conditions determining whether a particular ion can be discharged or not can be found by measuring the separate potential at each electrode during the electrolysis. Instead of observing the voltage across the whole cell, the potentials of the electrodes are obtained individually by connecting each with a reference e.g., calomel electrode. The variation of electrode potential with current is depicted by a curve similar to that shown before, the point D now corresponds to the discharge potential or deposition potential at the particular electrode.

If an electrolyte contains a number of different ionic species then, provided they are not disturbing factors, each ionic discharge will occur when the appropriate potential is reached. Since the process at a cathode involves taking up of electrons, i.e., reduction in the most general sense, cathodic processes will take place in the order of decreasing reduction potentials, that is in the order of increasing oxidation potentials. Similarly, anodic processes will occur in the order of decreasing oxidation potentials. In an acid solution of cupric sulphate, for example, there are present cupric and hydrogen ions, the oxidation potentials are – 0.34 and 0.0 volt, respectively, and so in accordance with the rule given above, cupric ions will be discharged first at the cathode. If the anode in the same solution is of copper then, three anodic processes are possible, namely dissolution of copper to form copper ions, discharge of sulphate ions and discharge of hydroxyl ions. The oxidation potentials for these processes in the given solution are about – 0.34, – 2.0 and – 1.2 volts respectively and hence the copper anode will dissolve in preference to the other processes. With a platinum anode, however, the corresponding oxidation potential has a large negative value and so there will be a discharge of hydroxyl ions accompanied by the evolution of oxygen.

It is frequently stated that in the electrolysis of an alkali hydroxide solution, the alkali metal is first discharged and the free atom attacks the water to give hydrogen; thus

$$Na^+ + e^- \rightleftharpoons Na$$

$$Na + H_2O \rightleftharpoons NaOH + \tfrac{1}{2} H_2$$

Since the sodium hydroxide is completely ionized and hence is equivalent to Na^+ and OH^-, the mechanism postulated is equivalent to saying that the sodium ions are not discharged and that the hydrogen comes from water. Similarly, the view that SO_4^{2-} ions are discharged at the anode in the electrolysis of aqueous sulphuric acid solution and that the sulphate radicals attack water, does not take into account the fact that the latter reaction namely,

$$SO_4^{2-} + H_2O \rightleftharpoons \tfrac{1}{2} O_2 + H_2SO_4 \rightleftharpoons 2H^+ + SO_4^{2-}$$

proves that sulphate ions are reluctant to be discharged. LeBlanc and others have expressed the opinion that hydrogen and hydroxyl ions are always discharged discretly, at cathode and anode respectively in solutions of both acids and bases.

In view of the above discussion, it may appear surprising that the decomposition voltages of solutions of alkali and alkaline earth salts of strong acids, e.g., $NaNO_3$, K_2SO_4 etc. are about 2.2 volts; if direct hydrogen and hydroxyl ion discharge occurs as suggested, the voltage should be 1.7 Volts. The explanation is that the discharge of these ions leaves an excess of hydroxyl and hydrogen ions in the vicinity of the cathode and the anode and since the solutions are not buffered they become alkaline and acid respectively. Hydrogen is thus liberated from an alkaline solution and oxygen from an acid electrolyte and this must introduce an additional voltage.

EXPERIMENT 77
Polarisation Phenomenon

Aim: To study the polarization phenomenon during the electrolytic discharge of a metal ion and to determine the value of exchange current density (i_o) as well as the charge transfer coefficient (α).

Apparatus Required: Potentiometric set-up, saturated calomel electrode, polarizing circuit, ammeter and copper electrode.

Chemicals Required: Copper sulphate solution (1 M) in H_2SO_4 medium.

Principle: When a metal electrode is in contact with its ions in solution the reversible potential is given by

$$E_{rev} = E^o + (RT/nF) \ln [M^{n+}]$$

where the reaction considered is $M^{n+} + n\,e^- \longrightarrow M$. However, when the metal electrode is sustaining a steady current density i (amp/cm^2) due to the deposition electrochemical reaction, the potential E of the electrode is different from the reversible potential value and depends on the magnitude of i. From the curves of the dependence of the elctrode potential on i, one can calculate the kinetic parameters of electrochemical reactions namely, the exchange current density i_o and the charge transfer coefficient α making use of the Tafel equation in the form

$$(i/i_o)\ \alpha\ \exp\,(-\alpha n f \eta/RT)$$

where η is the over potential given by $E - E_{rev}$. This equation is valid only when the reverse reaction is taking place to a negligible extent.

Procedure: Clean the electrodes with emery paper, water, hot 10% NaOH, dilute HC1 and water. Prepare a copper sulphate solution (1M) in dilute sulphuric acid medium. Immerse the electrodes and connect it to the polarizing circuit (Fig. 41). Now record the potential across a SCE and the polarizing electrode as a function of current density i, increasing i upto 0.05 amp/cm^2. Calculate the area of the electrode 'A' that is immersed in the solution

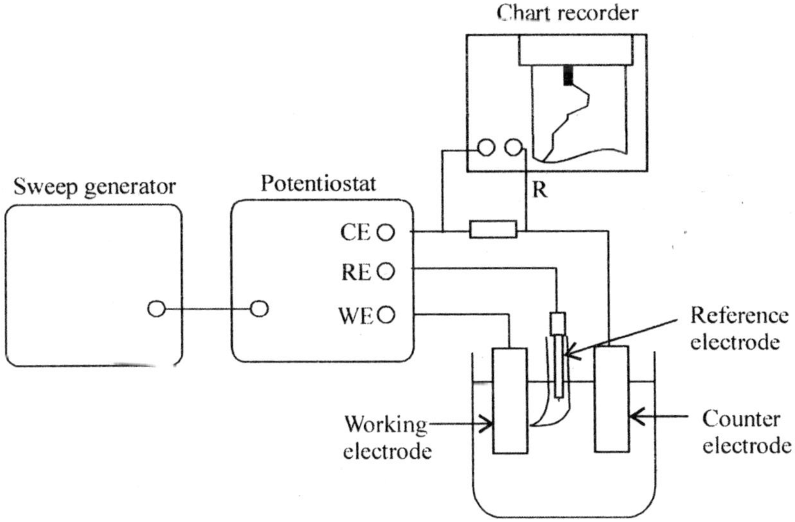

Fig. 41. measurement of polarization curve with swept potential (Polarising circuit).

Treatment of Data:

Tabulate the results as shown below

E (volt)	i (amp)	(Current density), i/A Amp/cm^3

1. Calculate the E_{rev} value for Cu^{2+} /Cu half cells corresponding to the concentration of the Cu^{2+} solution taken.
2. Plot the polarization curve, i.e., i versus E.
3. Make use of Tafel equation in the form where no mass transfer control is present namely,

$$i/i_o = \exp [-\alpha nF\eta/RT]$$

Calculate the value of i_o and α from intercept and slope.

Result:

Report the values of (i) exchange current density, $i_{o,}$ and (ii) charge transfer coefficient, α.

PART XI
Voltametry

EXPERIMENT 78

Reversibility of an Electrochemical Reaction and Determination of Concentration of a Given Reducible Ion—Polarography

Aim: To test the reversibility of an electrochemical reaction using dropping mercury electrode and to determine the concentration of the given reducible ion.

Apparatus Required: Polarographic set-up, calomel electrode, inert gas source.

Chemicals Required: Cadmium sulphate (about 0.01 M) in 1 M KCl.

Principle: When a reducible ion like Cd^{2+} is present in solution with excess of KCl, the current—potential curve for the electrochemical reduction of Cd^{2+} occurring at the freshly formed dropping mercury electrode shows a plateau, when diffusion is the only mode of transport for the Cd^{2+} ions to the electrode. The limiting current (in this case called limiting diffusion current) is given by the Ilkovic equation,

$$I_d = 607\ n\ C\ D^{1/2} m^{2/3} t^{1/6}$$

where C is the concentration of the reducible substance (in this case Cd^{2+} ions in millimoles/litre), D is the value of the diffusion coefficient ($cm^2\ sec^{-1}$), n is the number of electrons involved in the electrochemical reduction process, t is the drop time of mercury in seconds, and m is the mass rate of flow of mercury in mg/sec. This equation forms the basis for the estimation of the concentration of any reducible ions if all other parameters are kept constant. When one considers a kinetically reversible electrode reaction like

$$Cd^{2+} + 2\ e^- \rightleftharpoons Cd\ (Hg)$$

taking place at the dropping mercury electrode, the current - potential curve is described by the equation,

$$E = E_{1/2} + (RT/nF) \ln \{(I_d - I)/I\}$$

where I_d is the diffusion current, and $E_{1/s}$ is the half wave potential corresponding to $i = I_d / 2$. According to this equation, if the process is reversible then a plot of $\log \left(\frac{I_d - I}{I}\right)$ versus the applied potential should be a straight line with a slope whose magnitude is theoretically obtainable from the knowledge of the number of electrons involved in the electrochemical process.

In normal circumstances, a typical polarogram shows various regions and these are shown in (Fig. 42). The diffusion current is estimated as shown in the figure as the difference between the limiting current and the extrapolation of the residual or base current (the base or residual current is due to charging and other effects).

Procedure: Take a known quantity of cadmium sulphate in KCl in a beaker or electrochemical cell with provision for inserting the dropping mercury electrode as well as for bubbling pure nitrogen to remove dissolved oxygen. The electrochemical cell is coupled with a reference electrode, usually calomel electrode. Connect the calomel electrode to the positive end of the polarographic unit while the dropping mercury electrode is connected to the negative end of the polarographic unit. Record the polarogram by increasing the applied potential in steps of 25 millivolts or 50 millivolts, every time noting the current in the cell, upto one volt.

Treatment of Results:

1. Correct the current readings for the residual current.
2. Plot $\log \{(I_d - I)/I\}$ versus the applied potential.
3. Calculate the slope and compare with the theoretical value of the slope.
4. Identify the value of $E_{1/2}$ as the potential at $I_d/2$ value.

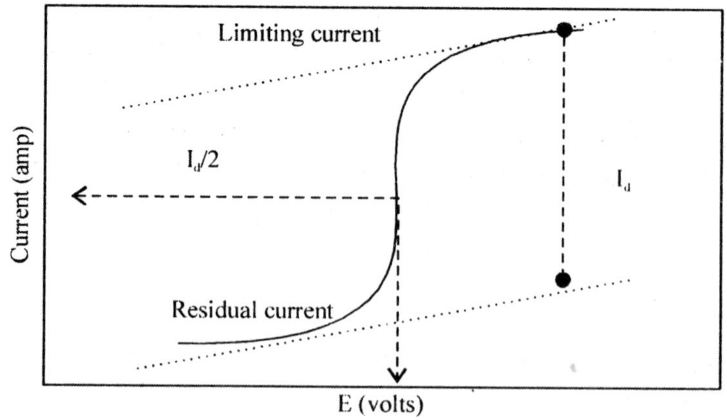

Fig. 42. Typical polarogram and measurement of diffusion current.

Results:

1. Report the value of $E_{1/2}$ and identify from the value of $E_{1/2}$ the nature of the species undergoing reduction.
2. Comment on the reversibility of the process.
3. Carry out the experiment at different concentrations of the reducing species and identify from I_d value, the concentration.

EXPERIMENT 79
Formula and the Stability Constant of a Complex by Polarography

Aim: To determine the stability constant of the lead oxalate complex by polarographic method.

Apparatus Required: Polarographic set-up, calomel electrode and inert gas source.

Chemicals Required: Potassium oxalate , 1.0 M; lead nitrate, 0.02 M and potassium nitrate, 1.0 M.

Principle: This is based on the principles of polarogrphy that when a steadily increasing voltage is applied to a cell containing an electroactive substance, incorporating a relatively large quiescent mercury electrode and a minute mercury cathode composed of a succession of small mercury drops falling slowly from a fine capillary tube, it is possible to construct a reproducible current - voltage curve. From an examination of the current-voltage curve, information as to the nature and concentration may be obtained.

Consider the dissociation of a complex ion

$$MX_p^{(n-pb)+} = M^{n+} + pX^{-b} \quad (1)$$

The instability constant may be written

$$K_{instability} = \{[M^{n+}]\,[X^{-b}]^p\}/\{MX_p^{(n-pb)+}]\} \quad (2)$$

We may imagine the amalgam reaction to be (assuming amalgam formation)

$$M^{n+} + ne^- + Hg \rightleftharpoons M\,(Hg) \quad (3)$$

Combining (1) and (3) we have

$$MX_p^{(n-pb)+} + ne^- + Hg \rightleftharpoons M(Hg) + pX^{-b} \quad (1)$$

It can be shown that the expression for electrode potential can be written

$$E_{1/2} = E_0 + (0.0591/n)\log K_{instability} - (0.0591/n)\log [X^{-b}]^p$$

Here p is the coordination number of the complex ion formed. X^{-b} is the complexing agent and n is the number of electrons involved in the reaction. The concentration of the complex ion does not enter into the above equation, so that the observed half wave potential will be constant and independent of the concentration of the complex metal ion. Furthermore, the half-wave potential is more negative, the smaller the value of $K_{instability}$ i.e., the more stable the complex ion. The half wave potential will also shift with a change in the concentration of the complex forming agent and if the former is determined at two different concentrations of the complex forming agent, we have

$$\Delta E_{1/2} = -(0.0591/n)\ \Delta\log\ [X^{-b}]$$

The relationship enables one to determine the coordination number *p* of the complex ions and thus its formula.

Procedure:

1. Pipette out 2 cm^3 of lead nitrate solution into each of the five 100 cm^3 volumetric flasks.
2. Add the following solutions into each of the flasks as instructed below.

Flask 1 : 10 cm^3 of 1.0 M potassium nitrate
Flask 2 : 10 cm^3 of 1.0 M potassium nitrate + 10 cm^3 of 1.0 M potassium oxalate.
Flask 3 : 10 cm^3 of 1.0 M potassium nitrate + 20 cm^3 of 1.0 M potassium oxalate.
Flask 4 : 10 cm^3 of 1.0 M potassium nitrate + 40 cm^3 of 1.0 M potassium oxalate.
Flask 5 : 10 cm^3 of 1.0 M potassium nitrate + 80 cm^3 of 1.0 M potassium oxalate.

Dilute each of these solutions to the mark and mix thoroughly. Flush nitrogen through each solution for 10 minutes.

Make the necessary connections in the circuit.

Apply the voltage gradually increasing it from zero to the required value in small steps and note the corresponding values of current (deflection observed on the galvanometer since it is proportional to the current) in the table given below. Using the data :

1. Draw a current voltage curve for each solution (see Fig. 42).
2. Measure the half wave potential $E_{1/2}$.
3. Plot $E_{1/2}$ against log $[(C_2O_4)^{2-}]$.

Slope = 0.0591 (p/n) and intercept = 0.0591 log (k/n) where n is the number of electrons involved in the reaction. Hence determine the stability constant K and p, the number of oxalate ions in the lead oxalate complex.

Treatment of Results:

Volume of $K_2C_2O_4H_2O$ cm^3 →	Current reading (µA)					
	0	10	20	40	80	
Voltage applied (V) ↓	1	II	III	IV	V	
0						
0.1						
0.2						
0.3						
0.4						
0.5						
0.6						
0.7						
0.8						
0.9						
1.0						
1.1						
1.2						
1.3						

Half wave potential ($E_{1/2}$) of lead = volts.

Volume of $K_2C_2O_4\ H_2O$	Concentration of $K_2C_2O_4\ H_2O$	Log x	$(E_{1/2})_C$ (volts)	$(E_{1/2})_C - (E_{1/2})_S$ (volts)

$E_{1/2}\ /_C$ half wave potential of the complex ion

Slope = . . . Therefore, the value of p is : . . .

Intercept = . . . Therefore, the value of K is: . . .

Results:

The stability constant of the lead oxalate complex = . . .

The number of oxalate ions in the lead oxalate complex = . . .

Hence, formula of the complex is . . .

EXPERIMENT 80
Reversibility of a Redox Process and Determination of the Concentration of a Given Solution — Cyclic Voltametry

Aim: To test the reversibility of a reduction-oxidation process as well as to monitor the concentration of the given solution.

Chemicals Required: 1 millimolar $K_4Fe(CN)_6$ in 0.1 M KNO_3 solution. (For better result make the solution just before the commencement of the experiment.)

Apparatus Required: Cyclic voltammetric unit, electrochemical cell shown in the (Fig. 43) with provision for three electrode system and provision for bubbling inert gas as well as stirring arrangement, suitable recorder, glassy carbon working electrode, Ag/AgCl reference electrode and platinum counter electrode.

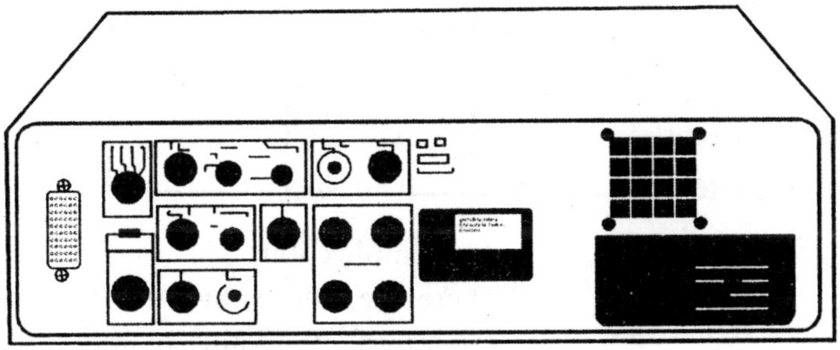

Fig. 43. Cyclic voltammetric unit.

Principle: This experiment is based on the principles of cyclic voltammetry in which the potential is scanned linearly with respect to time from an initial value E_i to a second value and then back to E_i or some other potential depending on the nature of the species undergoing oxidation or reduction. This potential excitation signal (with reference to a reference electrode normally with respect to SCE) can be applied repeatedly and hence the term cyclic voltammetry. The current response signal when the potential excitation is applied in a cyclic manner is shown in Fig. 44. As the potential is scanned positively anodic current occurs when the electrode becomes a strong oxidant to oxidize a reduced species like Fe^{2+}. The anodic current increases rapidly until the concentration of the reduced species (for example Fe^{2+}) at the electrode surface decreases to zero causing the current to peak. The current then decays at a rate proportional to $t^{-1/2}$ as the solution surrounding the electrode is depleted of the reduced species. In essence, the oxidized species is electrogenerated in the forward scan (anodic sweep) and the electrogenerated oxidized species is reduced back as indicated by the cathodic current. Thus cyclic voltammetry is capable of generating a new species during the forward anodic scan and then monitor its fate in the reverse cathodic scan.

The scan direction is switched to the negative direction for the reverse scan. When the potential is such that the electrode becomes a strong reductant, the oxidized species which is accumulated at the electrode is now reduced. Thus causes the cathodic current which increases until the surface concentration becomes zero. The current then peaks and decays as the concentration of the oxidized species is depleted adjacent to the electrode. The cycle is completed when the potential reaches a value of 0.0 volt.

Procedure: Take about 50 cm^3 of the solution in the electrochemical cell. Pass nitrogen for 15 minutes so as to remove any dissolved oxygen. Connect the three electrodes to the suitable positions of the cyclic voltammetric unit. Set the positions for the scan as follows:

Intial E	=	– 0.20 volt.
Positive (potential) limit	=	+ 0.80 volt.
Negative (potential) limit	=	– 0.40 volt.
Potential scan rate	=	20 mV per second.

Operate the instrument in the scan mode with suitable current amplification of the current signal. The scan will continue from – 0.20 volt until it reaches the positive limit of 0.80 volt where the potential scan is reversed. The scan is then continued in the negative direction until the negative limit of – 0.40 volt is reached. On the reverse sweep, a cathodic current should be observed which is due to the reduction of the ferricyanide generated on the initial sweep in the positive direction. A typical cyclic voltammogram for the ferrocyanide/ferricyanide system is shown in Fig. 44.

Treatment of Results:

1. Measure the cathodic and anodic currents at the peak positions.
2. Tabulate the data as shown in the table.

Scan rate millio volt/sec.	$V^{1/2}$ $(V/s)^{1/2}$	i_{pa} (μA)	i_{pc} (μA)	E_{pc}/E_{pa}	E_{pc} (V)	E_{pa} (V)	E_p (V)

3. Conclude on the reversibility of the process.
4. Calculate the number of electrons involved in the electrochemical reaction.
5. Plot Ip_a versus $V^{1/2}$ and note whether the variation is linear or not.

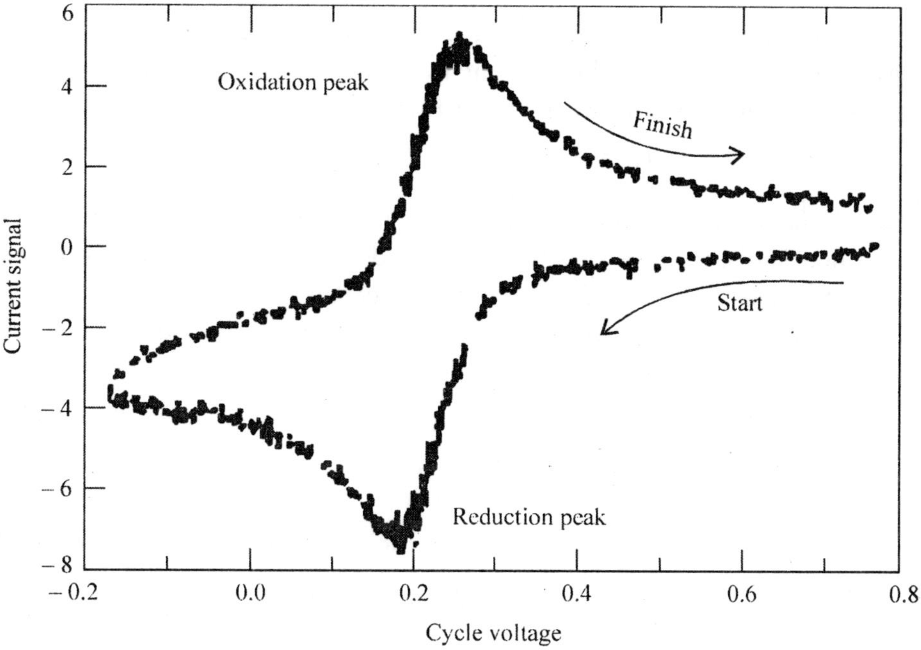

Fig. 44. A typical cyclic voltammogram showing reduction and oxidation current peaks.

Results:

1. Comment on the reversibility of the porcess.
2. Comment on the number of electrons involved in the electrochemical reaction.

Note:

The important parameters in cyclic voltammogram are:

1. The magnitude of the peak anodic current (i_{p_a}) as well as peak cathodic current (i_{p_c})
2. The anodic and cathodic peak potential (E_{p_a} and E_{p_c})

The method of measuring the current values is illustrated in the figure by extrapolation of the base line current. The number of electrons (*n*) transferred for a reversible couple is determined from the separation between the peak potentials, thus

$$E_p = E_{p_a} - E_{p_c} = 0.059/n$$

For one electron process the value of E_p should be 59 millivolt. The peak current for a reversible system is described by the Randles–Sevcik equation for the forward sweep in the first cycle

$$i_p = 2.69 \times 10^5 \, n^{3/2} A D^{1/2} C V^{1/2}$$

where i_p is the peak current in amperes, n is the electron stoichiometry (equivalent/mol), A is the area of the electrode (cm^2), C is the concentration (mol/cm^3) and V is scan rate in volts per second.

According to this equation i_p is proportional to $V^{1/2}$ and is directly proporational to concentration.

For a reversible couple i_{p_a} is approximately equal to i_{p_c} or $i_{p_a}/i_{p_c} \approx 1$.

The applications of cyclic voltammetry are:

1. Characterization of a redox system.
2. Determination of the formal potential (reduction) which is given by

$$E^0 = (E_{p_a} + E_{p_c})/2$$

(Caution: this is not true in the case of slow electron transfer process)

3. Since the fate of the electrogenerated species is being monitored during the reverse scan, it is possible to elucidate the mechanisms of electrochemical reactions as well as to identify any intermediates that are formed in the electrochemical reduction or oxidation.

PART XII
Miscellaneous Experiments

EXPERIMENT 81

Magnetic Susceptibility by Guoy Balance

Aim: To determine the magnetic susceptibility of a given substance like $FeCl_2$, and Fe_2SO_4 in powder form.

Apparatus Required: The Guoy balance, the samples, glass tube, DC power supply.

The Guoy Balance: The magnetic susceptibility is a phenomenon that arises when a magnetic moment is induced in an substance by the presence of an external magnetic field. Magnetic susceptibility, χ is the ratio of the magnetization, M (magnetic moment per unit volume) to the applied magnetic field, H. The magnetic moment can be measured either by force methods, which involve the measurement of the force, exerted on the sample by an inhomogeneous magnetic field or induction methods where the voltage induced in an electrical circuit is measured by varying magnetic moment. (The details of induction method will be given in a separate experiment). In force method (which are usually classified as Guoy method or Faraday method), the apparent change in the weight of the sample due to the application of a magnetic field, is measured by a sensitive balance and this weight change is directly proportional to the magnetic susceptibility, χ. In the Guoy method, a long sample is placed such that part of the sample is in the field and part, outside the field ($H = 0$). The magnetic field gradient is obtained by using truncated pole-pieces as shown in (Fig. 45). In the Faraday method the sample is placed in between two specially shaped pole pieces of a magnet which gives rise to uniform field gradient.

Thus the induced magnetic moment translates to a change in the weight of the substance when placed in an external magnetic field. This induced moment may have two orientations: parallel to the external magnetic field or perpendicular to the external magnetic field. The former is known as paramagnetism and the latter is known as diamagnetism. The physical effect of paramagnetism is an attraction to the source of magnetism (increase in weight when measured by Guoy balance) and the physical effect of diamagnetism is a repulsion from the source of magnetic field (decrease in weight when measured by a Guoy balance).

When a magnetic dipole of moment m is placed in a magnetic field H, the energy changes by an amount,

$$E = \frac{1}{2}\mu_0 mH$$

or

$$E = \frac{1}{2}\mu_0 VMH$$

where M is the magnetization and V is the volume of the specimen.

If the energy varies along the coordinate x, then the specimen experiences a force F, equal to

$$F = \frac{dE}{dx} = \frac{1}{2}\mu_0 V\frac{d(MH)}{dx}$$

For a diamagnetic or a paramagnetic meterial, M is proportional to H

$$F = \frac{1}{2}\mu_0 V\chi\frac{d(H^2)}{dx} = \mu_0 V\chi H\frac{dH}{dx}$$

And for a permanent magnet in a relatively small field, we can assume M to be constant and in such a case,

$$F = \frac{1}{2}\mu_0 VM\frac{dH}{dx}$$

Thus in both cases, it is necessary to have a non-uniform field to produce the force. As the specimen has a finite volume, it is necessary that the force acting on different parts of its volume should be the same. In the case of diamagnetic or paramagnetic materials, we must make sure that d (H^2)/dx is uniform over the volume of the specimen and in the case of material with a constant magnetization, dH/dx must be uniform. In both cases, the field itself may be parallel or perpendicular to x: the force is in the direction along which the field varies, not in the direction of the field itself. The non-uniform field may be produced in two different ways. One is to use a solenoid or electromagnet to produce initially a uniform field. A set of coils is then placed in the field to produce an additional non-uniform field. The second way is to use an electromagnet with non-parallel pole faces. An easier version of the force method is to use the specimen in the form of a long rod/a specimen filled in a long non-magnetic tube. When the sample is placed in between the poles of an electromagnet, the field acting on the farther end of the sample will be small. The force experienced by the sample is proportional to the square of the magnetic field and thus the magnetic susceptibility.

Procedure: In the experimental set up shown in (Fig. 45), NS is an electromagnet energized by a DC power supply. The variable magnetic field is provided by the wedge-shaped pole-pieces. The entire electromagnet is housed inside a wooden casing. The distance between the pole-pieces can be varied by means of a handle on top of the wooden casing. A digital balance is placed which carries a hook at the bottom for suspending the glass tube containing the material ($FeCl_2$, or Fe_2SO_4). The magnetic field between the pole pieces can be varied by changing the current through the coils using a DC power supply.

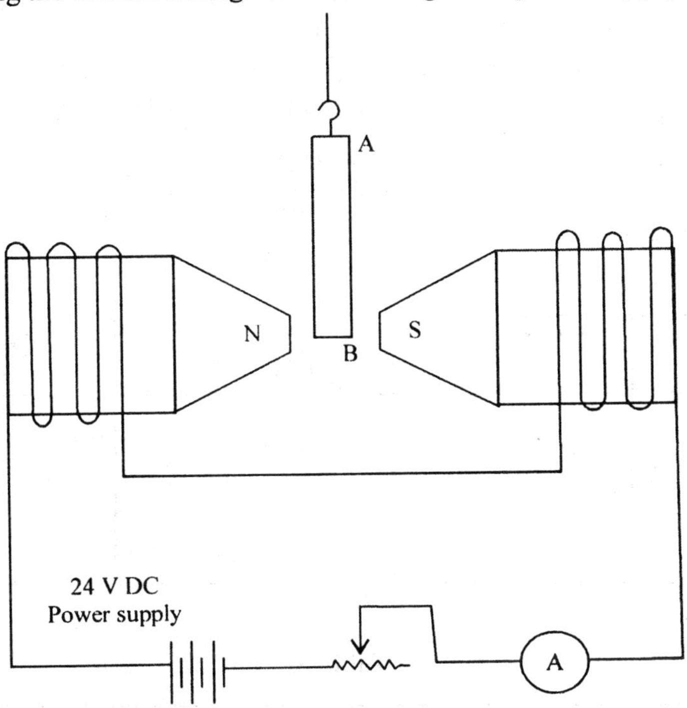

Fig. 45. Guoy Balance. [NS – Electromagnet with a power supply, AB – glass tube with a hook to be suspended to a chemical balance.]

A calibrated chart of the current vs. the magnetic field between the pole-pieces can be obtained.

1. First adjust Zero of the digital balance.
2. Suspend the empty glass tube as shown in (Fig. 45) and find its weight in zero magnetic field.

3. Using the D.C. power supply, vary the current from 0 to 3.5 A in steps of 0.2 A and in each case find the weight of the glass tube, i.e. determining the weight of the glass tube at various applied magnetic fields. The change in weight may be very small as the susceptibility of air is $\simeq 0$.
4. Now take out the glass tube and fill in the given substance (say $FeCl_2$) to about 3/4th of the tube. [If the sample is a solid, grind it thoroughly and pack it tightly into the sample tube] Find the weight of the glass tube with substance with an accuracy of 10 mg., in zero magnetic field. As before, find the weight of the substance in different applied magnetic fields (both for the increasing and decreasing fields). While adjusting the magnetic field by varying the current value, wait for a couple of minutes for the field to stabilize, and take the reading.
5. Repeat the above steps as necessary.

Note: Carefully reduce the current through the magnetic coils and reduce to zero and then only switch off the power supply.

Treatment of Data:

Table I.

S. No.	Wt. of empty glass tube (g)	Current through the coils (A)	Magnetic field (Gauss)

Table II.

S. No.	Wt. of the substance (g)	Current through the coils (A)	Magnetic field (Gauss)

Formula:

$$mg = \frac{A}{2}(\chi - \chi_o)H^2 \tag{1}$$

where m is the mass of the substance, A is the area of cross-section of the glass tube, H the magnetic field in between the pole-pieces, χ magnetic susceptibility of the substance, χ_o magnetic susceptibility of air.

The magnetic susceptibility of air is almost zero then relation 1 reduces to

$$mg = \frac{A}{2}\chi H^2 \tag{2}$$

A graph is drawn between m and H^2 and the slope gives $A\chi/2g$. Hence the susceptibility, χ is calculated. Also calculate molar susceptibility.

Result:

Report the value of the susceptibility of the sample.

EXPERIMENT 82

Transition Temperature and Heat of Transition by Differential Thermal Analysis

Aim: To determine the transition temperature and the heat of transition of the given sample like $CuSO_4.5\ H_2O$ or $BaTiO_3$-$PbTiO_3$.

Apparatus Required: Differential thermal analyzer, potentiometer for temperature measurement, amplified–potentiometer or micro-voltmeter for differential temperature measurement (all these may be part of the custom made differential thermal analyzer [DTA] equipment).

Chemicals Required: Reference sample either calcined alumina or $PbZrO_3$ and the substance under investigation, $CuSO_4\ 5\ H_2O$ or the solid solution powder sample.

Principle: The differential thermal analysis is a technique wherein the temperature differences between a sample and standard reference material (usually calcined alumina) are measured as both the samples are heated or cooled. When there is no heat change either in the sample or the reference, the temperature of both the substance will be the same and there will be no signal from the differential thermocouple. On the other hand when the sample undergoes a phase change or undergoes a reaction, it absorbs or evolves heat energy depending on whether the process is endothermic or exothermic. In that case the substance appears to be cooler or warmer than the standard and the temperature difference between the two appearing as the differential thermal EMF. The shape of the signal in the form of a peak indicates the total heat change accompanying the process and the maximum of the peak indicates the temperature at which the process is centered. Thus the area under the curve in the ΔT versus T plot is proportional to the heat of the transition or process taking place in the sample.

The ideal reference standard is so chosen that (i) it does not undergo phase transitions in the temperature range of interest so that there is no ambiguity of the origin of the observed heat effects; (ii) it has nearly the same heat capacity as the sample thus minimizing the drift of the differential EMF and (iii) it has the nearly the same density of the sample in order that the volumes of the sample and reference are almost the same. This experiment can be carried out not only for the determination of phase transitions but also a variety of other reactions wherein there is a definite thermal energy change in the process.

These include

(i) Thermal decomposition reactions of solids like $CaCO_3$, $CuSO_4.5H_2O$ and a variety of other materials like clays, hydrotalcites.

(ii) Phase transitions in a variety of systems.

(iii) Glass transition temperatures and synthesis of materials.

Procedure: Weigh accurately about one gram of the standard sample and place it in the stainless steel crucible in the DTA unit. Weigh about one gram of the sample and place it in the other stainless steel crucible in the DTA apparatus. Make sure the sensing thermocouples are centrally located in the two crucibles and are in thermal equilibrium with the crucibles.

Now heat the furnace at a given rate of heating say 10°C per minute upto 700°C. Record the temperature of the sample and the differential temperature using suitable amplifier and indicator.

Now turn off the furnace and also measure ΔT versus T while the furnace is cooling to room temperature.

Plot ΔT versus T on a graph paper.

Determine the temperature at which the peak has a maximum.

Also determine the area under the peak and utilize it to determine the heat of the reaction by the use of a standard substance whose reaction heat is known. Make sure that the standard substance also undergoes reaction near the temperature of the reaction of the substance under investigation.

Results:

Report the temperature of transition.

Report the heat of transition.

Indicate the errors involved and account for the same.

EXPERIMENT 83
Polymorphic Transformation of Solids — Dilatometry

Aim: To study the polymorphic transformation in solid materials by dilatometry.

Apparatus Required: Fused silica dial gauge dilatometer, temperature controller, furnace with power supply, potentiometer, chromel-alumel thermocouple, samples about 10 × 10 × 100 mm size.

Principle: Phase transitions in solids can be divided into two kinds, namely reconstructive where the primary bonds are broken and new primary bonds are formed or displacive where bond lengths and bond angles alone are changed. For example silica can exhibit both kinds of phase transformations. The length changes accompanying the displacive transformations can be measured through a dilatometer. When these changes are substantial, as in quartz or other materials the substance develops cracks on repeated heating and cooling through the transformation temperature. For example zirconia undergoes a monoclinic-tetragonal phase transition at 1170°C accompanied by a disruptive volume change. The transition temperature can be lowered by adding about 10 percent CaO and eliminated by the introduction of 15 percent CaO. Zirconia containing 15 percent of CaO has cubic symmetry and does not exhibit any phase changes. This material hence is called *stabilized Zirconia*.

Many ferroelectric materials such as $BaTiO_3$, $PbNb_2O_6$ and antiferro electric materials like $PbZrO_3$ exhibit displacive phase changes which have a drastic effect on dielectric properties. These phase changes can be detected and studied with the help of a dilatometer.

For example iron transforms from alpha form (bcc) to gamma form (fcc) at 1183 K and then goes over to the delta form (bcc) at 1673 K.The melting point of iron is 1812 K. Transformations in steel and graphitization in cast iron are accompanied by volume changes and can therefore be detected by dilatometric methods. The kinetics of these transformations may be studied by measuring the length changes as a function of time at fixed temperature in the transition range.

Procedure: Assemble the dilatometer as shown in Fig 46. Measure the temperature and dial guage readings on one of the samples. Record the temperature of the sample at every stage. At the initial temperature (room temperature) record the length of the specimen in mm and note it as L_0.

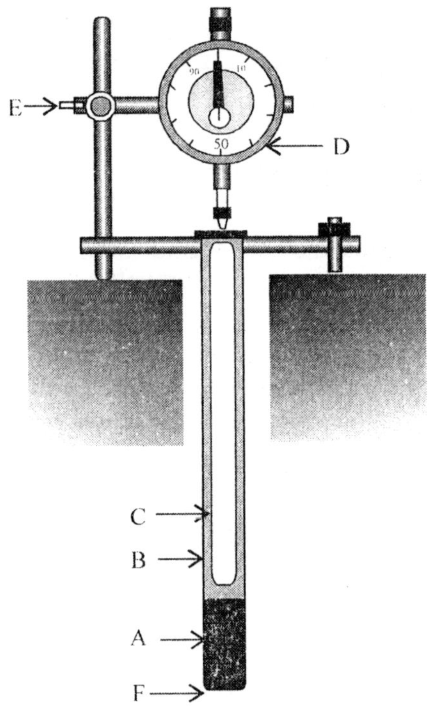

Fig. 46. Dilatometer assembly.

Treatment of Data:

Tablulate the results as given below

EMF	Temperature (K)	Dial Gauge reading	ΔL (mm)	$\Delta L/L_0$

Calculate and comment on the following:

1. Calculate the coefficient of linear thermal expansion, which is given by

$$\alpha = (1/L_0)\ (\Delta L/\Delta T)$$

2. Calculate the total length change at the transformation and assuming isotropy calculate the percent volume change.
3. Compare the transformation temperature with those given in literature for the sample used.
4. Do you see any hysteresis in the heating and cooling cycles? Is the hysteresis observed is a due to effects caused by the instrument or due to the sample. If hysteresis is due to the sample explain the effect observed.
5. Calculate the density of the high temperature polymorph making use of the data of density at room temperature and the thermal expansion behaviour of the sample.
6. Repeat the experiment with small impurities in the sample and comment on the effect of the impurity on the polymorphic change of the material.

Result:

Report the results obtained for each of the observations.

EXPERIMENT 84
Electrochemical Nature of Corrosion and its Control

Aim: To demonstrate the electrochemical nature of aqueous corrosion and also to study the methods of corrosion control.

Apparatus Required: DC millivoltmeter, DC milliammeter, Pyrex beakers, emery paper for cleaning metals and wires , wires and clips for making connections , variable resistors (low wattage) and dry cell.

Chemicals Required: 3 % saline solution, Dilute hydrochloric acid (0.2 N) for cleaning the metal surfaces, Strips or rods of copper, brass, zinc, iron, steel aluminum, nickel, graphite and various other metals, if available.

Principle: Normally it is considered that aqueous corrosion is electrochemical in nature. If a piece of iron is immersed in a polar solvent like water some of it will dissolve. Iron dissolved in water is always in the form of ions, never as neutral atoms. Thus the dissolution reaction is written as

$$Fe \rightleftharpoons Fe^{++} + 2e^{-}$$

which denotes the dissolution of the metal as metal ions if the forward direction of the reaction is considered and in the reverse direction it is the plating of the metal from the solution. However, the dissolution cannot go on endlessly since it is a equilibrium limited dissolution process. In order to demonstrate the electrochemical nature of the corrosion process let us consider two metals iron and copper in electrical contact with each other. When iron goes into solution the electrode is retaining the negative charges which can be transported to the copper electrode where it can combine with the metal ions to deposit the metal. So in this Galvanic cell mode the iron acts as the anode and copper functions as the cathode. Now connect the copper to the positive end and iron to the negative end of a voltmeter and measure the open circuit voltage. The points to be noted are: 1. There are many metals like magnesium and zinc which have tendency to dissolve in water stronger than that of iron. In these cases these metals act as anodes and iron acts as the cathode. Similarly when platinum and copper is used Platinum functions as the cathode and copper as the anode. It is clear that the tendency to dissolve depends on the nature of the electrolyte. Therefore the Galvanic potential measured for copper and iron will be different for pure water, sea water and acid solution. Now imagine that we close the circuit for the iron and copper containing Galvanic cell. In this case one will observe that dissolution takes place from the anode and deposition takes place in the cathode. This implies both ionic and electronic current must flow in the internal circuit and external circuit respectively. Since ions are leaving the anode, the anode will undergo changes and corrosion occurs at the anode and since these ions are deposited in the neutral state at the cathode, no corrosion will occur at the cathode. The current flowing is an indication of the corrosion taking place and one also measure the extent of corrosion from the weight changes of the anode. Since at the cathode other reduction reactions can also occur like the evolution of hydrogen from the acidic solution or the reduction of oxygen to give rise to OH^- ions according to the equation

$$O_2 + 2H_2O + 4e^- \longrightarrow 4OH^-,$$

The rate of corrosion reaction depends on the surface area of the cathode, as well as the cathode surface conditions to promote one or the other reactions considered.

Procedure:

1. Setting up of a Galvanic cell and measuring its potential:

Take about 250 ml of the 3 % saline water in a dry beaker. Place a strip of aluminum and one strip of Zinc in the solution. Note the magnitude of the potential developed and the polarity of the electrodes. Now replace Aluminum by Copper. The magnitude of the voltage developed should be higher. Note the value and account for the higher value observed. Also note the potential developed as a function of the separation of the electrodes and also the geometry of the electrodes employed. Note if there is any change in potential as a result of cleaning the electrodes.

2. Galvanic series:

In the above experiment conduct and tabulate the voltages developed using a series of electrodes copper, aluminum, iron, steel, nickel, brass and other metals against each other. See if you can arrange the potential values in any consistent manner.

3. Cathode polarization:

Construct a galvanic cell using copper (cathode) and zinc (anode) electrodes after cleaning the respective metallic strips. Measure the potential as a function of time at room temperature. Repeat this experiment with different areas of the electrodes. Make sure every time when you start the measurement after cleaning the electrodes. Find the relationship between corrosion rate and the area of the electrode.

4. Sacrificial anode:

Construct a cell using iron (anode) and copper (cathode). Note the current and the potential. This will ensure that the corrosion is taking place in the solution. Now place another electrode say zinc and connect the zinc electrode electrically to the iron electrode. Now note the corrosion rate by measuring the potential and current between the iron and copper electrodes.

Another way to develop a sacrificial electrode is to use a scrap iron which is connected to the anode iron electrode though an accumulator. Slowly increase the potential applied to the sacrificial electrode and note the potential at which the iron corrosion current measured between iron and copper electrode used is minimum.

Result:

Report your observations for each of the above four steps.

EXPERIMENT 85
Current — Voltage Characteristics of a p-n Junction Rectifier

Aim: To study the current-voltage characteristics of a p – n junction rectifier.

Apparatus Required: Semiconductor diode or equivalent, 12 volt dry cell, potentiometer, press switch DPDT switch, voltmeter, ammeter various ranges 0 - 100 mA and 0 - 50 microamperes range.

Principle: When a p type semiconductor and an n type semiconductor are in contact their Fermi levels (close to the valence band in the case of p type semiconductor and close to conduction band in the case of n type semiconductor) equalize as shown in (Fig. 47.) This gives rise to a contact potential difference V_0.

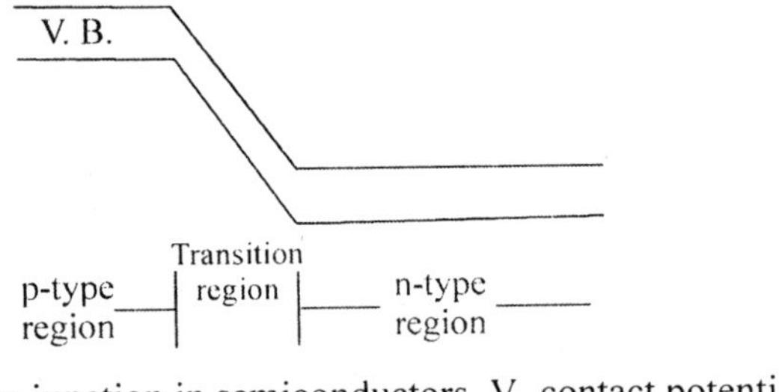

Fig. 47. p - n junction in semiconductors. V_0 contact potential difference.

When a forward bias is applied, the p-side has a positive voltage with respect to the n-side thus reducing V_0 and opposite is the case in the reverse bias. These situations are shown in Fig. 48.

For the forward bias the current is given by $I = I_0 [\exp(eV/kT) - 1]$

And, since V is greater than kT/e + 0.025 volt $I = I_0 \exp(eV/kT)$

For the reverse bias $I = I_o$ [exp $(-eV/kT) - 1$], Or $I = -I_o$.

Here I_o is the equilibrium value of the current due to electrons (in this case minority carriers from the p-side) flowing to the right in Fig. 48.

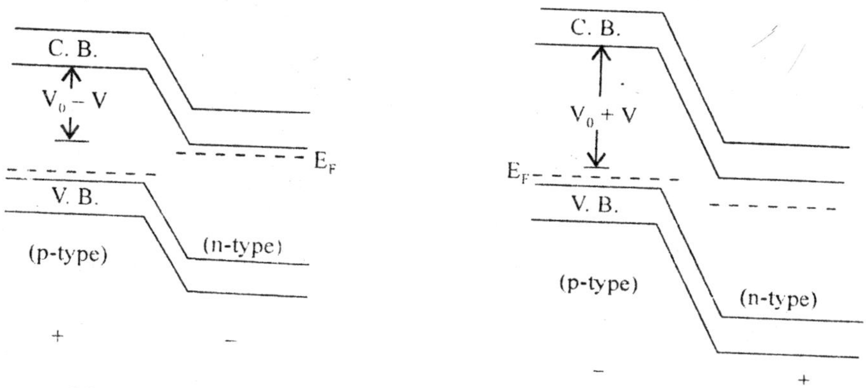

(a) Forward bias. (b) Reverse bias.

Fig. 48. p-n junction with external bias voltage applied.

Typical current voltage curve obtained for this rectifying action is shown in (Fig. 49).

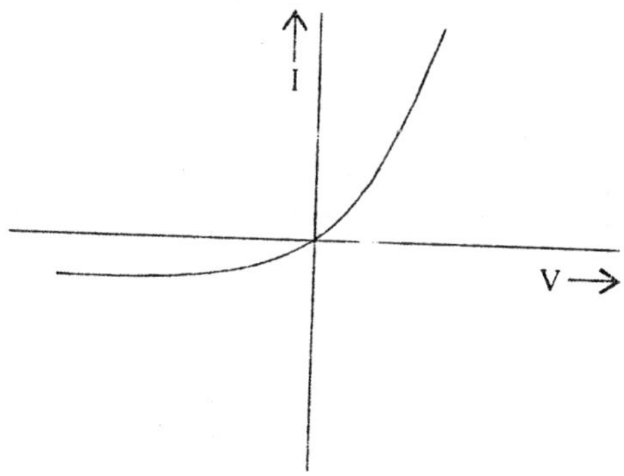

Fig. 49. I-V characteristic for p-n junction.

Procedure: The circuit arrangement used for this measurement is given in Fig. 50. By pressing switch 1 the power is supplied to the circuit. The DPDT switch in position 1 and 2 connects the diode for direct bias and reverse bias measuments. The potential applied to the diode can be varied continuously by means of the potentiometer P.

Keep the switch S_2 in position 1. Press S_1 and apply 0.5 volt as read in the voltmeter. Note the current in ammeter for every voltage readings. Now turn switch S_2 to position 2 and repeat the measurements.

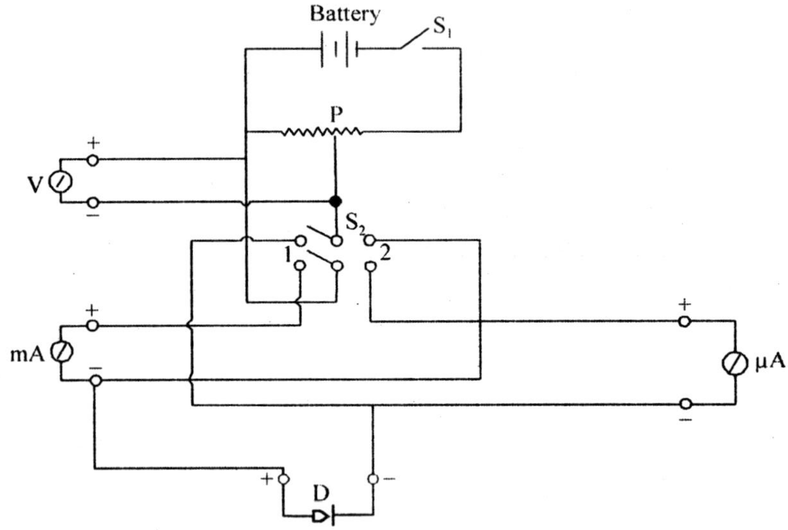

Fig. 50. Circuit diagram for I - V characteristic measurement in a p - n junction diode.

Treatment of Data:

Tabulate the results as follows:

Bias condition	Voltage (V)	Current measured (I) (amps)
Forward bias	0.50	
	0.75	
	1.00	
	1.25	
	1.50	
	1.75	
	2.00	
	2.25	
	2.50	
	2.75	
	3.00	
Reverse bias	1.0	
	2.0	
	3.0	
	4.0	
	5.0	
	6.0	

Bias condition	Voltage (V)	Current measured (I) (amps)
Reverse bias	7.0 8.0 9.0 10.0	

Results:

Report the following:

1. The value of the saturation current I_0 in the reverse bias direction.
2. Calculate the effective resistance of the diode as a function of voltage.

EXPERIMENT 86
Dielectric Behaviour of Ceramics by Phase Transition

Aim: To study the dielectric behaviour of ceramic barium titanate through the tetragonal–cubic transformation.

Apparatus Required: Ceramic disk of barium titanate, micrometer, sample holder, a furnace, chromel alumel thermocouple, potentiometer, capacitance bridge.

Chemicals Required: Barium titanate.

Principle: Barium Titanate, $BaTiO_3$ exhibits the following polymorphic transformations

$$\text{Rhombohedral} \xleftrightarrow{-80^\circ C} \text{orthorhombic} \xleftrightarrow{0^\circ C} \text{tetragonal} \xleftrightarrow{120^\circ C} \text{cubic}$$

in this experiment we shall study the tetragonal to cubic transition. The non centro- symmetric location of the titanium ion sets up a permanent dipole and causes spontaneous ionic polarization. When the direction of the spontaneous polarization of a material can be reversed by an applied electric field, the material is called a *ferroelectric* and the phenomenon, *ferro electricity*.

Procedure: Measure the diameter and thickness of the sample using the micrometer. Record the readings.

Place the sample disk in the specimen holder and the specimen holder in the furnace. Connect the specimen holder to the capacitance bridge. Measure the capacitance and dissipation factor at a frequency of 1 kc/s at room temperature and at 10 degree interval upto 150°C. Take readings at closer intervals of temperature in the vicinity of the phase change.

Also repeat the measurements during the cooling cycle.

Observations

Thickness of the specimen = mm

Diameter of the electrode =

Area of the electrode =

EMF (mV)	Temperature (°C)	Capacitance	ε	Dissipation factor

$$\varepsilon = C/C_0 = Cd/(A\ \varepsilon_0)$$

where C is the measured capacitance in farads, C_0 is the geometrical capacitance in vacuo, d is the thickness of the capacitor in meters, A is the electrode area of the capacitor in square meters. ε_0 is the permittivity of vacuum (8.854×10^{-12} farad/m).

Results:

Record the following points

1. What is the temperature coefficient of capacitance in the temperature interval 300 K till 354 K?
2. How does the observed value compare with the reported value of dielectric constant? Account for the difference.
3. Report the variation of dissipation factor around the transition temperature.

Note:

A ferroelectric material exhibits the following characteristics:

1. High dielectric constant 10-10,000 compared to 1-10 for most materials.
2. A hysteresis loop between the dielectric displacement D and the electric field E.
3. A Curie temperature beyond which the material becomes paraelectric losing its spontaneous polarization. And above Curie temperature the Curie Weiss temperature dependence is observed.
4. A sharp peak in the dielectric constant versus temperature plot occurring at the Curie temperature.

EXPERIMENT 87
Partial Molal Volume of Ethanol — Water Mixture

Aim: To determine the partial molar volume of ethanol – water mixture at a given composition.

Apparatus Required: Specific gravity bottle, pipettes, ten stoppered bottles, weighing box and balance.

Chemicals Required: Absolute ethanol and pure water.

Principle: Partial molar volume is a thermodynamic quantity which reflects on the ion-solvent interactions. The extensive properties like volume, energy are not truly additive when pure components are mixed. For example when one mole of water is mixed with one mole of ethanol the resultant volume of the mixture is not the additive of the volumes of the independent components. Let us illustrate this with more details.

Benzene and toluene form an ideal solution. The volume of 1 mole pure benzene is 88.9 ml; the volume of 1 mole pure toluene is 106.4 ml. According to equation

$$V = n_1V_1 + n_2V_2$$

88.9 ml benzene mixed with 106.4 ml toluene results in 88.9 ml + 106.4 ml, or 195.3 ml of solution.

Common sense suggests that the volumes add up, since the volumes of substances in solution are extensive properties. On the other hand, water and ethanol do not form an ideal solution. The volume of 1 mole pure ethanol is 58.0 ml and the volume of 1 mole pure water is 18.0 ml. However, 1 mole water mixed with 1 mole ethanol does not result in 58.0 ml + 18.0 ml, or 76.0 ml, but rather 74.3. Therefore it is known that the partial molal volumes are additive or extensive properties. When the mole fraction is 0.5, the partial molal volume of ethanol is 57.4 ml and the partial molal volume of water is 16.9 ml. One can now calculate the volume of the solution:

$$1 \text{ mole} \times 57.4 \text{ ml/mol} + 1 \text{ mole} \times 16.9 \text{ ml/mol} = 74.3 \text{ ml}$$

which is exactly what is observed. Note that the values just cited for the partial molal volumes of ethanol and water are only for a particular concentration; in this case, the mole fraction is equal to 0.5 and applies only to the water-ethanol system.

Therefore the partial molal volume of a component can be defined as *the change in volume of the mixture by the addition of one mole of the component at constant temperature, pressure and moles of the other component.* Therefore the partial molal volume can be defined as

$$V_1 = (\delta v/\delta n_1)_{T,P,n_2}$$

and

$$V_2 = (\delta v/\delta n_2)_{T,P,n_1}$$

where n_1 and n_2 are the number of moles of the two components.

The total volume of the mixture is given by the equation

$$V = n_1 V_1 + n_2 V_2$$

It is therefore clear that ideal solutions obey Raoult's Law over the entire range of compositions and hence there is no change in volume or energy content at all temperatures and pressures. Systems which show negative deviation from Raoult's law show decrease in volume on mixing the two components. This type of deviation is shown by water-ethanol mixtures.

Procedure: Wash and dry the bottles thoroughly. Prepare a series of solutions of ethanol and water say 2 + 18 mL, 4 + 16 mL, ... 18 + 2 mL in the given stoppered bottles. Determine the density of pure ethanol, water and each of the solutions by using a density bottle or pyknometer.

Treatment of Data: Molal volume of ethanol and water in the mixtures can be calculated using the following equations:

V_m(ethanol) = Molecular weight of ethanol/density of that mixture
= 46.07/density mL/mol.

V_m(water) = Molecular weight of water/density of that mixture
= 18/density mL/mol.

Now calculate the specific volume (reciprocal of density) of pure ethanol water and each of the solutions. Plot percentage composition of mixture by volume (abscissa) against specific volume i.e. $1/\rho$ (ordinate). Draw a smooth curve through the points and draw tangents to the curve at specified compositions of the mixture. Calculate the partial molar volumes. Then plot partial molar volume against composition and show the variation of partial molar volume with concentration.

Results:

Comment on the variation of partial molar volume with composition.

Comment whether the total volume is greater or less than the sum of the volumes of the pure components.

Are the partial molal volumes of the solute and solvent greater or less than the pure molal volumes? Comment on the nature of interaction.

EXPERIMENT 88
Velocity of Sound in Liquids — Ultrasonic Interferometry

Aim: To find the velocity of sound in liquids using ultrasonic interferometer.

Apparatus Required: Quartz crystal oscillator, resistors, microammeter, liquids.

Principle: Longitudinal waves with frequencies above 20 KHz are termed as ultrasonic waves. A piezo-electric crystal slab (like Quartz) with parallel faces can develop charges of opposite polarity on their faces when it is strained. This causes an electrostriction (some mechanical strain) in the crystal slab, when placed in a region of electric field. When a high frequency oscillatory voltage is applied to the parallel faces a quartz crystal slab, the slab produces acoustic vibrations of exactly the same frequency. Thus ultrasonic vibrations of high frequencies (of the order of a few mega hertz) can be produced by connecting the parallel faces of a quartz crystal slab (electrical contact being provided by thin metallic films coated on the faces) to a high frequency electronic oscillator. Standing wave pattern due to superposition of ultrasonic waves can be formed in liquid. The distance between any two consecutive nodes represents half the wave length ($\lambda / 2$) of the ultrasonic waves. If f is the frequency of the vibrations, then the velocity of propagation of the ultrasonic wave in the liquid is

$$V = f \cdot 2d \qquad (1)$$

The experiment therefore deals with the method of measuring the distance between consecutive nodes.

Procedure: The ultrasonic interferometer is shown in (Fig. 51). The procedure for the experimental investigation is given in the following steps:

STEP 1 Find the least count of the screw head attached to the reflector.

STEP 2 Fill the cell with the experimental liquid (say water) so that the cell is 3/4th full. Fix the reflector unit on top of the cell. Mount the cell in the given sturdy metal base. Connect the output cable from the ultrasonic apparatus to the socket provided on the metal base.

STEP 3 Switch on the ultrasonic apparatus. Keep the adjust and gain knobs in the minimum positions. Just after a minute or so, the meter reading will register a maximum. Wait for some more time (i.e. about a minute or so). The meter needle will come back to zero reading on the dial.

STEP 4 Turn the gain knob to a maximum. The reading in the meter may be around 50 or 60 microamperes. Now turn the adjust knob so that the meter reading is around 80 microampere.

STEP 5 Turn the screw gauge head such that the pitch scale reading is the least. In this position, the reflector is closest to the crystal – the source of ultrasonic vibrations.

Let the screw be turned and the reflector moved up. The meter reading in the ultrasonic apparatus will change. Note the Head scale as well as the Pitch scale reading (on the screw gauge) when the microammeter reading is the least.

STEP 6 Repeat the experiment and note down the readings of the Pitch scale and the Head scale for successive minima readings in the microammeter.

Treatment of Data: Tabulate the readings.

Number of minimum reading of the microammeter	Screw gauge reading			Distance between 10 consecutive minima
	PSR	HSR	Actual reading	

Mean =

STEP 7 The mean value of the readings of the last column of the tabulation will represent λ/2 of the wave from which V can be evaluated. Using the formula

$$V = f\lambda$$

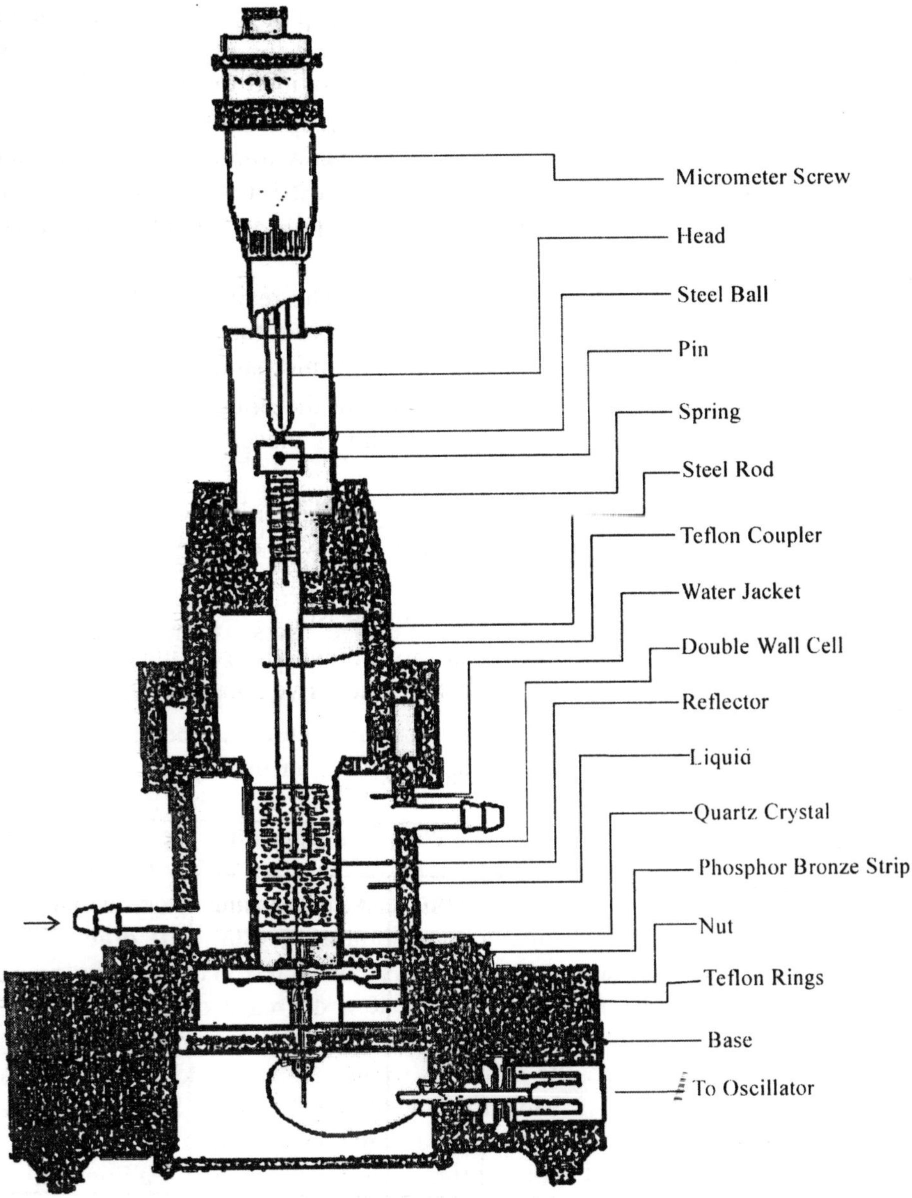

Fig. 51. Interferometer Cell

the velocity V of ultrasonic wave propagation in the liquid can be evaluated. The frequency f of the oscillation has to be obtained from the data sheet of the unit in MHz.

STEP 8 Remove the liquid from the cell. Pour a small quantity of another liquid, in which the velocity is to be determined. Rinse the cell with that liquid. Pour it out. Then fill the cell with that liquid and carry out the experiment as described in the previous steps.

STEP 9 Repeat the experiment for different liquids provided.

These experiments can be carried out at various temperatures, or in a mixture of liquids or that of a salt solution. In the latter two cases if the compositions are known then unknown composition can also be determined.

Result:

Report the value of the speed of sound in the liquids used and also compare with the values reported in literature.

Note:

The ultrasonic interferometer set up has basically a Wheatstone's network. Two arms of the network are the load resistances of two identical pentode oscillators, the plate resistance of the pentodes forming the other two arms. The oscillations have been stabilized using a 2 MHz Quartz crystal oscillator. A microammeter is connected between the junctions of the load resistance and the radio frequency choke. This acts as the detector. The output of one of the oscillators is fed to a 2 MHz Quartcrystal mounted to a cell. The cell can be filled up with any desired liquid in which the velocity of sound is to be determined. A reflector can be moved up or down using a screw head. The metal reflector can be kept as close to the crystal transducer as possible. A finite current may be made to pass through the microammeter. This is done by adjusting the 'Adjust' and 'Gain' Knobs. The reflector, when moved, can go to such a height, a standing wave pattern can be formed between the transducer crystal (source) and the reflector. At this moment, the vibrating crystal will draw energy and thus load the oscillator to which it is connected. The loading causes a greater imbalance in the bridge leading to a decrease of current through the microammeter. The meter thus gives a minimum reading. As the reflector is moved further, the standing wave pattern could not be formed and the meter needle goes back to its original position. Thus, whenever the reflector is at such a distance from the crystal (source) forming a standing wave pattern and representing a node, the meter reading will be a minimum. The distance 'd' moved by the reflector to register two consecutive minimum readings in the microammeter will correspond to (λ /2) of the ultrasonic wave in the liquid. Thus a measurement of 'd' and a knowledge of the frequency 'f' of the oscillator will enable one to determine the velocity 'v' of propagation of the ultrasonic waves in the given liquid medium. Measurements can also be made on the the variation of velocity of ultrasonic waves with temperature.

EXPERIMENT 89

Determination of Very Low Concentration of Metals in Solution by Flame Photometry

Aim: To determine the concentration of sodium, potassium, lithium and calcium ions in the given samples.

Apparatus Required: Flame photometer, standard flasks and pipettes.

Chemicals Required: Standard solutions (0.01 M) of sodium chloride, or potassium chloride, or calcium chloride and the solution of the given substance suitably prepared with known dilution.

Principle: When an aqueous solution of a salt is introduced or aspired into a flame, the following events occur rapidly.

1. The solvent, in this case water is evaporated away leaving behind the solid salt.
2. The salt is vapourized into the gaseous state.
3. The gaseous molecules are dissociated into free neutral atoms.
4. Some of these neutral atoms are then excited by the thermal energy of the flame.
5. These excited neutral atoms emit radiation and return to the ground state.

The emission spectra is characteristic line spectra with appropriate wavelength corresponding to the energy of the emitted radiation of the neutral atoms which were excited. The intensity of the emitted radiation is proportional to the amount of elements present in the sample. The intensity of radiation is dependent on the temperature of the flame. However in the case of alkali metals increase of temperature may give rise to ionization instead of excitation. Hence temperature control is necessary which can be controlled by suitable fuel-oxidant composition. If a series of known concentration solutions are prepared and the intensities of emitted radiations are measured one can generate a calibration curve which can be used to determine the concentration of the unknown solution.

Procedure: Prepare 0.01 M KCl solution in deionized water. This solution will contain 390 ppm of potassium ions. From this stock solution prepare solution containing 100 ppm of potassium ions. Then prepare 25, 20, 15, 10, 8, 6, 4, 2 ppm solutions from the stock solution. Now set up the flame photometer by igniting the flame and allow the flame to reach steady state. Make sure the flame has a nice blue colour by suitably adjusting the air and fuel control knobs. Place the potassium filter in the flame photometer. Then fill up the beaker of the flame photometer with deionized water and dip the outer end of the inlet tube inside the beaker. As the water is aspirated into the flame adjust the galvanometer to the zero of the reading. Now empty the beaker and pour the 25 ppm solution and aspirate the solution in the flame. When the galvanometer shows steady reading, adjust the reading to 100 by adjusting the knobs. Repeat these two adjustments for setting 0 and 100 of the galvanometer reading by repeatedly aspirating pure water and 25 ppm solution. Now introduce the 20 ppm solution in the beaker and aspirate it to the flame. Record the reading. Again introduce pure water and then introduce 15 ppm solution and note the galvanometer reading. Thus record the galvanometer reading for each of the standard solutions. Introduce the given unknown solution and note the reading.

Similarly one can repeat this experiment from stock solutions of other ions using appropriate filter.

Treatment of Data:

[Salt] (ppm)	2	4	6	8	10	15	25	50	Unknown
Meter reading									

Draw a calibration curve by plotting concentration of salt (in ppm) on x-axis against the galvanometer readings (on y-axis). Calculate the concentration of the given unknown solution in ppm from the calibration chart by interpolation. Repeat the procedure for other ions.

Result:

Report the concentration of the respective ions in the unknown solution.

Note:

Make sure the type of filters one uses for the estimation of each of the metal ions and wavelength range where the emissions occur.

Find out the accuracy of the determination and comment on the magnitude.

EXPERIMENT 90

Specific and Molar Refraction of a Liquid by Refractometry

Aim: To determine the specific and molar refraction of a given liquid.

Apparatus Required: Abbe refractometer, dropper, cotton wool and density bottles.

Chemicals Required: Acetone, methyl acetate, ethyl acetate, n-hexane and carbon tetra chloride.

Principle: The refractive index (n) is defined as n = (sin i /sin r) where i and r are the angles of incidence and refraction. The refractive index decreases with increase in temperature but the specific refraction or specific refractivity defined by the following equation is independent of temperature.

$$R_s = (n^2 - 1)/(n^2 + 2) \times 1/\rho$$

Where R_s is the specific refraction, n is the refractive index and ρ is the density of the liquid. Molecular refraction is defined as

$$[R] = (n^2 - 1)/(n^2 + 2) \times (M/\rho)$$

where M is the molecular weight of the substance. The molecular refraction is an additive and constitutive property. Each atom in a compound has a definite contribution to the molar refraction of a compound.

The molar refraction of a homogeneous solution is the sum of the molar refractions of its constituents. So one can write

$$[R]_{mixture} = X_1 [R_1] + X_2 [R_2]$$

X_1 and X_2 are the mole fractions and $[R_1]$ and $[R_2]$ are th molar refractions of the components of the mixture.

Molar refraction of the mixture can also be determined by the formula

$$[R]_{mixture} = [(n^2_{mix} - 1)/(n^2_{mix} - + 2)] [(X_1M_1 + X_2M_2)/\rho_{mix}]$$

where n_{mix} and ρ_{mix} are the refractive index and density of the mixture or solution.

For the quantitative analysis of binary solutions the following expression can be used

$$R_3 = R_1 (p/1000) + R_2 [(100 - p)/100]$$

where R_1, R_2 and R_3 are the specific refractions of components 1, 2 and the mixture respectively and p is the percentage of component 1.

Procedure: Place the refractometer on a table near a window so that sufficient light reaches the prism. Open the prism box by turning the lock nut. Clean the faces of the both prisms with the help of cotton wool, alcohol and acetone and close the prism box after drying. Introduce few drops of liquid in the prism through the small hole on the prism box, by means of a dropper such that a film of liquid will be enclosed between the two prisms. Focus the cross wire of the telescope by rotating the eye piece and adjust the mirror to reflect maximum light into the prism. Move the prism box backward and forward until a clear boundary between the light and dark regions appear. Now rotate the prism box until the sharp boundary line coincides with the intersection of the cross wire in the telescope. Read the refractive index directly on the scale. Also determine the density of the liquid.

Treatment of Data:

Enter the refractive index values of the liquids or solutions used.

Liquid				
Refractive index, n				

Calculate the values of specific refraction and molar refraction values of the given substance, using the respective equations given in the principle section above.

Result:

Report the values of the specific refraction and molar refraction of the liquids. Compare with the values reported in literature. Account for the differences observed, if any.

***Note*:**

We can also extend this experiment to

(i) determine the concentration of the salt in a given unknown solution,

(ii) determine the atomic refractions for a given substance

(iii) assess the contribution to the molar refraction by some typical groups like CH_2.

To determine the percentage composition of a given mixture: If you are given two liquids find out the refractive indices of the pure liquid and also that of the unknown mixture given. Determine the density of the pure liquids and that of the mixture. Substitute the values obtained in the expression given in the principle section and determine the composition of the given mixture.

Report the composition of the given mixture and compare with the correct value from the instructor.

To determine the concentration of a salt in a solution: Weigh exactly 2, 4, 6, 8, 10 and 12 grams of pure KCl and dissolve in distilled water separately in 100 ml standard flasks and dilute to the mark. Determine the refractive index of each of the solution and also that of the unknown solution.

Plot a calibration graph between concentration and refractive index and determine the concentration of the unknown solution.

To determine the atom refraction : Determine the refractive index of each of the liquids methyl acetate, ethyl acetate, n-hexane and carbon tetra chloride. Also determine the densities of the above liquids using density bottle. Calculate the molar refractions of the liquids. From these values calculate the following:

1. [R] of ethyl acetate – [R] of methyl acetate = refraction of a CH_2 group.
2. [R] for n-hexane – 6 x refraction of $CH_2(C_6H_{12})$ = refraction of 2 H.
3. Refraction of CH_2 – 2 x refraction of H (2H) = refraction of C
4. [R] of CCl_4 – refraction of C = refraction of 4 Chlorine.

To determine the refraction of methylene group: Determine the refractive indices of ethyl acetate, propyl acetate and butyl acetate. Determine the densities of the liquids using density bottle. Calculate the molar refraction of each liquid. Treat the data as follows:

1. [R] of butyl acetate – [R] of propyl acetate = refraction of – CH_2
2. [R] of propyl acetate – [R] of ethyl acetate = refraction of – CH_2

Show that value of refraction of – CH_2 group is constant.

EXPERIMENT 91
Dipole Moment of a Liquid

Aim: To determine the dipole moment of a polar compound like acetophenone or methyl benzoate from measurements of the densities and dielectric constants of dilute solutions of the compound in cyclohexane.

Apparatus Required: A density measuring device (pycnometer), Abbe refractometer, impedance bridge, audio oscillator.

Chemicals Required: Acetophenone or methyl benzoate.

Procedure: Prepare solutions of 1.0, 2.0, 3.0, 4.0, 5.0 and 7.0 mole percent of the polar compound in cyclohexane. The densities of these solutions, and of pure cyclohexane, acetophenone or methyl benzoate, are to be determined with a pycnometer.

The refractive index of either acetophenone or methyl benzoate is determined with an Abbe refractometer using light from a sodium vapor lamp. The use of Abbe refractometer is described in Experiment 90. The refractive index should be measured for all molar solutions along with the pure substances. The dielectric constants of cyclohexane and the dilute solutions of the polar compound in cyclohexane are determined by the bridge method. The impedance bridge circuit is a generalized Wheatstone bridge with both capacitance and resistance in the measuring arms, an audio oscillator and a detector with an amplifier built in to increase the sensitivity. The impedance of an a.c. circuit can be considered as a generalization of the d.c. resistance and has components of resistance, inductance, and capacitance. For this experiment the inductance contributions are negligible.

The bridge is constructed so that the resistance component (called a dissipation factor) and the capacitance component of the unknown impedance are directly read off of the instrument

The operation of the bridge consists in adjusting the impedances Z_1, Z_2 and Z_3 (Z_4 is the unknown) until the voltage indicated on the detector is zero (or minimized), then,

$$Z_1/Z_2 = Z_3/Z_4$$

The oscillator and the amplifier are turned so as to balance the bridge . The oscillator is adjusted to a frequency of 5 KHz and the filter on the amplifier is tuned to the same frequency. The unknown is connected to the bridge and the gain control on the amplifier is adjusted so that the detector dial reads approximately 50 with the response set on logarithmic. Bridge Balance Procedures Fig. 51A or any other procedure applicable to the bridge being used can be adopted.

a. Move C MAX control to a setting slightly in excess of approximate value of capacitor being measured.

b. Set D MAX switch to 0.01 and set all DISSIPATION FACTOR controls to 0.

c. Position first (left-most) CAPACITANCE control for minimum deflection on detector meter.

d. To refine balance, position remaining CAPACITANCE controls, beginning at left and proceeding to right, until no further improvement occurs.

e. Manipulate DISSIPATION FACTOR controls, starting with right-most, to refine balance further.

f. Alternately manipulate first CAPACITANCE and then DISSIPATION FACTOR controls until final null is accomplished. An increase in D MAX may be required.

g. Observe direct CAPACITANCE readout in picofarads in the digital windows.

h. Observe direct DISSIPATION FACTOR reading, if appropriate, from digital windows.

   At 1 kc, dissipation factor is as indicated; otherwise, multiply by frequency in kilocycles.

i. If appropriate, observe CONDUCTANCE in ohms, reading from digital windows. Multiply reading by M factor (red engraving on C MAX dial) to calculate final value.

   The procedure is for a particular type of capacitance bridge but if any other bridge is used the instructions of the manufacturer of the capacitance bridge may be followed and the value of capacitance recorded.

j. With proper sequential adjustment of dissipation factor and capacitance, the detector reading will either increase or decrease when the lever-switch corresponding to the significant figure of the capacitance is moved one position. The bridge is then balanced and the capacitance is recorded.

k. The dissipation factor is not used in the calculations but it must be determined in order to balance the bridges.

l. If the capacitance reading for dielectric A between the plates of the capacitor is C (A), then the dielectric constant of A is given by:

$$\varepsilon = C\ (A)/C\ (air)$$

where C (air) is the capacitance when air is the dielectric between the plates. Once determined, C (air) is used to calculate the dielectric constant of all the liquids. The plates of the capacitor must be carefully rinsed with cyclohexane and air dried completely before each determination of C (liquid). A small amount of water can introduce enormous errors since its dielectric constant is about 30 times that of the solutions used. Hence care should be exercised not to allow any water contamination.Record the temperature at which the density and dielectric data are taken.

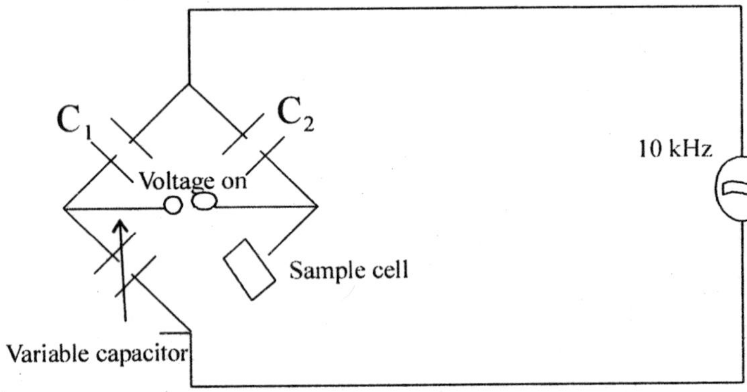

Fig. 52. Balance bridge for dipole moment measurement.

Calculations:

1. Graph the dielectric constants and densities of all the solutions as a function of mole fraction of the solute. Determine the slopes and intercepts of these graphs and their standard deviations by the least squares method.
2. Calculate P_0^2M, the limiting molar polarization of your polar compound, P_0^2d, the molar distortion polarization of your polar compound, and P_0^2m, the molar orientation polarization of the polar compound.
3. Determine the dipole moment of the polar compound.
4. Calculate the value of m (exp) in Debyes. Use vector addition and the table of bond moment data given below to calculate a value for m for the molecule under study. Make any assumptions regarding the coplanarity or lack of coplanarity of the various functional groups of the given molecule.

Bond Moments (Debyes, positive sign indicates moment is directed to second atom)

C – H – 0.40;	C – C 0.00 (definition)
C = O + 2.40 (in ester)	C – O + 0.74 (in ester, from acid carbon)
C = O + 2.30 (aldehyde)	C – O + 1.14 (in ester, from alkyl to oxygen of acid)

Use appropriate geometry of the molecule under consideration to compute the net dipole moment.

Result:

The dipole moment of the compound studied = ...

Note:

Pycnometers: Density Determination of Liquids.

Pycnometers are volume-calibrated containers used to determine the density of liquids or solids. They are particularly convenient for determining the density of liquids.

General Procedure: Record the mass of the empty pycnometer with the capillary stopper in place. Fill the pycnometer with the liquid of interest and put the capillary stopper in place; some of the liquid should overflow from the capillary. Using a Kimwipe, dry the outside of the pycnometer and then wipe the top of the capillary stopper clean. Record the mass of the filled pycnometer. To obtain the mass of the liquid, subtract the mass of the empty pycnometer from the mass of the filled pycnometer. Use the calibrated volume of the pycnometer and the mass of the liquid to determine the density of the liquid.

Calibration: Calibrate the pycnometer (accurately determine its volume) using distilled water as the standard (V = m/r). Use a handbook to get the density of distilled water at its measured temperature. Average at least three volume determinations.

***Note*:**

Use of pycnometer has been referred to in a number of places in this volume. Hence it is necessary to use these instructions at those places also. To avoid evaporation losses the measurement has to be carried out as quickly as possible.

Appendix II : Abbe Refractometer: Determination of the Index of Refraction. The use of this instrument is also discussed in a number of other experiments and the same can be used in this experiment also.

EXPEREMNT 92
Estimation of Avogadro Number

Aim: To determine the value of Avogadro's Number by electrolysis of water.

Apparatus Required: A DC source, ammeter, a bare copper electrode, an insulated copper electrode and three binder clips, a graduated cylinder (100 mL), a burette (50 mL) and a precision balance.

Chemcial Required: 1 M H_2SO_4.

Principle: Electrolysis uses the energy associated with moving electrons (electromotive force) to drive two chemical reactions – one at the anode, the other at the cathode. Electrons generated from an external source move onto the cathode causing it to become negative. Two nearby hydrogen ions (H^+) accept two electrons from the cathode and become H_2 – Hydrogen gas that bubbles out of the solution. At this point, a charge imbalance begins to form. The two H^+ ions had formerly served as counter-ions for two HSO_4^- ions (these were placed in the aqueous solution as H_2SO_4). Now an excess of negative ions (HSO_4^-) causes the energy of the system to increase. The reaction will cease if the energy increase of the system is not quickly addressed. To alleviate this buildup of energy, electrons must be given to the anode allowing the completion of the circuit and forming a new species. Two options exist for this new species: either it was negatively charged and became neutral, or it was neutral and became positively charged. In theory, two water molecules (H_2O) could be converted into O_2 and 4 H^+ ions effectively completing the electrolysis of water – H_2 generated at the cathode and O_2 at the anode. However, as in all chemical reactions, the pathway of lowest energy will be chosen. In the system devised for this laboratory, cations will come from the metal that serves as the anode. Specifically, copper atoms from the anode will give up two electrons each to form copper ions (Cu^{2+}). These ions will serve as counter-ions to the HSO_4^- ions mentioned previously. This process will continue as long as electrons move onto the cathode from an external source. The two half-cell reactions are shown below:

$$2H^{+}(aq) + 2e^{-} \longrightarrow H_2(g) \qquad E^{\circ} = 0.00\ V$$

$$Cu(s) \longrightarrow Cu^{2+}(aq) + 2\,e^{-} \qquad E^{\circ} = -0.34\ V$$

To determine Avogadro's Number, we need to determine how many electrons have traveled from the cathode to the anode and how many moles of electrons have traveled from the cathode to the anode. Then,

$$\text{Avogadro's Number} = \frac{\text{Number of electrons}}{\text{Moles of electrons}}$$

The numerator in the above equation is determined from the amperage and time.

Procedure: Follow the steps:

1. In a 100 mL graduated cylinder, place 30 mL of 1 M H_2SO_4.
2. In a 50 mL burette, add 1 M H_2SO_4 to the top of the burette.
3. Then, holding the graduated cylinder in your left hand at an angle and the middle of the burette in your right hand, invert the burette into the graduated cylinder. If more than 10 mL of air collects in the burette you will need to repeat this step.
4. Clamp the graduated cylinder in place.
5. Insert the curved bare end of the insulated copper electrode into the graduated cylinder and into the opening in the inverted burette.
6. Weigh the completely bare electrode to the nearest 0.001 g and place it into the graduated cylinder so that it contacts the sulfuric acid solution.
7. Use binder clips to connect the negative terminal of the DC source to the COM port of the ammeter and the other lead from the ammeter to the cathode.
8. Attach the positive terminal of the DC source to the anode with a binder clip.
9. Record the water level in the burette.
10. Record the temperature of the 1 M H_2SO_4.
11. Record the barometric pressure.
12. Plug the DC source into a 220 V outlet and start the stopwatch.
13. Allow the reaction to proceed until the hydrogen collected has forced the solution level in the burette to the 5 -mL mark.
14. Unplug the DC source and record the time.

15. Adjust the burette vertically until the level of liquid inside the burette is the same as the level of liquid in the graduated cylinder. Record the volume of Hydrogen collected.
16. Determine the pressure of the Hydrogen collected – recall that the pressure inside the burette is equal to the atmospheric or barometric pressure when the liquid levels are equivalent. Also, recall that this "total pressure" is equal to the $P_{H_2} + P_{H_2O}$ according to Dalton's Law of Partial Pressures. So, the P_{H_2} will be equal to $P_T - P_{H_2O}$.
17. Remove the bare electrode (anode) and make it dry with a paper towel. Weigh this electrode to the nearest 0.001 g.
18. Calculate Avogadro's Number from the Hydrogen gas data and the amperage and time.
19. Calculate Avogadro's Number from the loss of copper from the anode and the amperage and time.
20. Repeat the experiment.

Treatment of Data:

Record your results in the tabulation sheet.

Determination of Avogadro's Number

	Trial – 1	Trial – 2
1. Initial mass of the bare electrode	.... g	.... g
2. Final mass of the bare electrode	.... g	.... g
3. Mass of copper lost	.... g	.... g
4. Initial volume in burette	.... mL	.... mL
5. Final volume in burette	.... mL	.... mL
6. Volume of hydrogen collected	.... mL	.... mL
7. Temperature of hydrogen collected	.... °C	.... °C
8. Temperature hydrogen collected	K	.... K
9. Barometric pressure	.... mm Hg	.... mm Hg
10. Vapor pressure of water	.... mm Hg	.... mm Hg
11. Pressure of hydrogen collected	.... mm Hg	.... mm Hg
12. Time of the reaction (seconds)	.... sec	.... sec
13. Average current (amperes)	.... A	.... A

	Trial – 1	Trial – 2
14. Moles of copper lost from the anode (shown work below)	.... moles	.... moles
15. Moles of hydrogen formed (shown work below)	.... moles	.... moles
16. Copper lost from the anode (show work)		
17. Avogadro's number from moles of Hydrogen collected (shown work)		

Result:

The value of Avogadro Number evaluated from this experiment = ...

Alternate method:

This method is based on the fact that Avogadro's number is the number of atoms or molecules present in a sample with a mass, expressed in grams, numerically equal to its atomic or molecular mass expressed in amu's or Daltons. It is defined as the number of ^{12}C atoms in exactly 12 grams of ^{12}C.

Under appropriately controlled conditions, it is possible to spread a film of a fatty acid substance only one molecule thick, a monolayer, onto a water surface. The thickness of the monolayer can be calculated from the measured volume of substance required to make the monolayer and the measured area of the surface that is covered.

Substances such as stearic acid ($C_{17}H_{35}COOH$) are called fatty acids due to their presence in many natural fats. Such molecules include two major fragments: A long hydrocarbon chain and a carboxylic acid group.

Stearic Acid, $C_{17}H_{35}COOH$

$CH_3CH_2\,CH_2\,CH_2\,CH_2\,CH_2\,CH_2\,CH_2\,CH_2\,CH_2\,CH_2\,CH_2\,CH_2\,CH_2\,CH_2\,CH_2\,CH_2COOH$

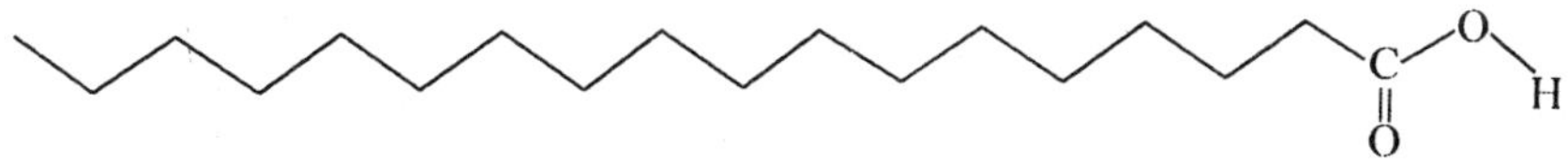

The polar-COOH acid group is attracted to water molecules whereas the nonpolar *hydrocarbon* chain is repelled by water molecules. Stearic acid is not soluble in water so to make a solution, it needs to be dissolved in a different solvent. When a solution of stearic acid in a volatile hydrocarbon solvent is placed on water, the stearic acid spreads out on the surface of the water such that each molecule has its polar –COOH group in the water and its nonpolar hydrocarbon chain above the water surface. The hydrocarbon solvent evaporates quickly, leaving behind a monolayer of stearic acid. Additional aliquots of stearic acid solution will cause the fatty acid molecules to spread and become more compact, adopting a vertical orientation until the entire water surface is covered. This is because each molecule "wants" to have its polar group in the water and the hydrocarbon chain in the air. The fully compact state of the monolayer is shown schematically in (Fig. 53).

Subsequent addition of stearic acid solution will cause it to float on top of the existing monolayer forming a lens-shaped droplet that is readily visible. The number of drops of stearic acid solution required to form the monolayer can thus be determined. The mass of stearic acid in the monolayer can be calculated from the measured volume and known concentration of the stearic acid solution. The number of moles of stearic acid in the monolayer is determined using its molar mass.

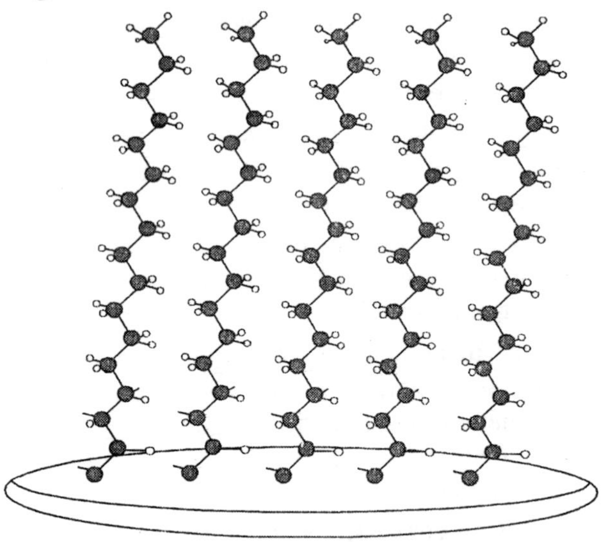

Fig. 53. Schematic magnified view of a stearic acid monolayer on a water surface. Only a few of the many molecules are shown for clarity.

We do not directly know the surface area of the water occupied by each stearic acid molecule. However, we can make guesses based on the molecular geometry of stearic acid. If we assume each stearic acid molecule on the surface has a cubical shape with side length s, it then has a height s and occupies an area s^2. The total number of stearic acid molecules in the monolayer can be determined from s^2 and the measured area of the monolayer. Avogadro's number is obtained by dividing the number of molecules of stearic acid in the monolayer by the number of moles of stearic acid in the layer.

Note:

Volt : a unit used to measure the potential difference between two points. The greater the potential difference (or voltage) between two points, the greater the tendency for the electrons to move between those two points.

Resistance : a force that retards the flow of electrons. Small wires have large resistances because it becomes increasingly more difficult to move larger and larger numbers of electrons through the wire in a given time. Large diameter wires have lower resistance and many more electrons can move along these wires in a given time. Resistance is measured in ohms.

Ampere : a measure of the number of electrons that pass a given point in a given time. An Ampere is equivalent to the voltage divided by the resistance—it is the tendency of electrons to move divided by the retarding force.

Coulomb : a measure of the electric charge that can be delivered. A large electric charge (or shock) can be delivered from high voltage, low resistance (large diameter) wires. The magnitude of the charge (or shock) increases in proportion to the length of time that the current flows. Therefore, a coulomb is equal to the ampere multiplied by the time.

Faraday : named after Michael Faraday who discovered the basic laws of electrolysis—is the magnitude of charge on one mole of electrons. It has been determined that the charge on one mole of electrons is equal to 96,500 Coulombs. This conversion factor allows one to convert coulombs to Faradays.

The following series of conversions will allow us to determine the numerator (number of electrons that traveled from the cathode to the anode)

$$\text{Amps} \times \text{Seconds} \times \frac{1\ \text{Coulomb}}{1\ \text{Amp - Sec}} \times \frac{1\ \text{Faraday}}{96500\ \text{Coulombs}}$$

$$\times \frac{1\ \text{Mole Electrons}}{1\ \text{Faraday}} \times \frac{6.02 \times 10^{23}\ \text{Electron}}{1\ \text{Mole Electrons}}$$

The denominator will be determined in two different ways allowing us to calculate Avogadro's Number from two different measurements. First, the volume of Hydrogen gas formed at the cathode can be measured and converted into moles of Hydrogen using the ideal gas equation PV = nRT. Then, the moles of electrons involved in the production of the Hydrogen gas can be calculated (2 moles of electrons are needed in the production of 1 mole of Hydrogen gas). Second, the loss in mass of the copper electrode can be measured and this mass converted into moles. Then, the moles of electrons involved in the conversion of copper metal into copper ions can be calculated (2 moles of electrons are required to convert 1 mole of copper metal into Cu^{2+} ions).

The amperes and seconds are converted into the number of electrons involved in the process. On the right side, the mL, temperature, and pressure of the collected Hydrogen is used to determine the moles of Hydrogen collected using PV = nRT. Conversion of moles of H_2 to moles of electrons provides all the information necessary to calculate Avogadro's Number.

Alternatively, the number of moles of electrons involved in the process will be determined by measuring the amount of copper loss from the anode. Conversion of moles of Cu to moles of electrons provides all the information necessary to calculate Avogadro's Number.

EXPERIMENT 93
Crystal Structure Determination — X-ray Diffractogram

Aim: To determine the structure of a crystal by X-ray Diffraction method.

Apparatus Required: X-ray diffractometer.

Chemicals Required: Any crystalline solid sample.

Principle: X-ray diffraction is one of the important characterization tools used in solid state chemistry and materials science. One can determine the size and the shape of the unit cell for any compound most easily using the diffraction of X-rays. The principle used in such analysis is based on the Bragg's equation,

$$n \times \lambda = 2\ d\ \sin\theta$$

where λ is the wave length of the incident X-ray used, θ – the angle of diffraction, n – the order of reflection and d is the interplanar distance. Making X-ray of known wavelength incident on the sample at different angles and measuring the intensity of the X-ray pattern at 2θ, the interplanar distance can be arrived at. From this the unit cell dimension and hence the structure of the solid can be worked out.

Procedure: Follow the steps.

1. Place the powder sample onto the double-side tape which is then placed on an aluminum sample holder; since it is a powder sample, use a spatula to spread the powder onto the double-side tape.
2. Read the instructions for the Miniflex X-ray diffractometer.
3. Set the instrument at optimum setting as follows: time constant 2; range (select the range of 2θ values to be covered and chart speed: Low).
4. Slide in the sample holder and adjust the beginning 2θ at 70 degree (It scans from high degrees to low degrees).

5. Switch on the start knob and chart recorder (slow) simultaneously, run your sample on slow chart speed.
6. Once scan gets down to 3° of 2θ, stop (switch start knob to off) and chart. Turn off X-ray.
7. Locate all peaks on the chart and corresponding 2θ values and write their values into the data chart below. Perform the necessary calculations in the table and calculate the repeat distance for the unit cell.

Treatment of Data:

Tabulate the data from X-ray diffractogram.

2θ	θ	sin θ	n	d = n × λ/sin θ	Index spacing n × d
			1 2 3 4 5 6 7 8		

Wavelength = 1.5418 Å for Cu Kα.

RESULT:

The interplanar distance d =

Note 1:

X-rays are electromagnetic radiation of wavelength about 1 Å (10^{-10} m), which is about the same size as an atom. They occur in that portion of the electromagnetic spectrum between gamma-rays and the ultraviolet. The discovery of X-rays in 1895 enabled scientists to probe crystalline structure at the atomic level. X-ray diffraction has been in use in two main areas, for the fingerprint characterization of crystalline materials and the determination of their structure. Each crystalline solid has its unique characteristic X–ray powder pattern which may be used as a "finger print" for its identification. Once the material has been identified, X-ray crystallography may be used to determine its structure, i.e. how the atoms are packed together in the crystalline state and what the inter-atomic distances and angles are.

The path difference between two waves: 2 × wavelength = 2d sin θ

For constructive interference between these waves, the path difference must be an integral number of wavelengths (i.e.,) n × wavelength = 2x

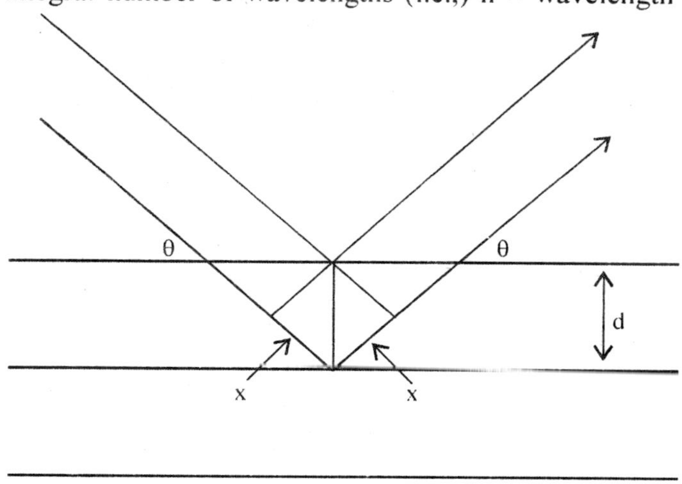

Fig. 54. Reflection of X-rays from two planes of atoms in a solid.

This leads to the Bragg equation:

$$n \times \lambda = 2\, d \sin \theta$$

Note 2:

To make the reader familiar with the X-ray pattern, the X-ray diffraction pattern from a single crystal of a layered clay is given below (Fig. 55). Strong intensities can be seen for a number of values of n; from each of these lines we can calculate the value of d, the interplanar spacing between the atoms in the crystal.

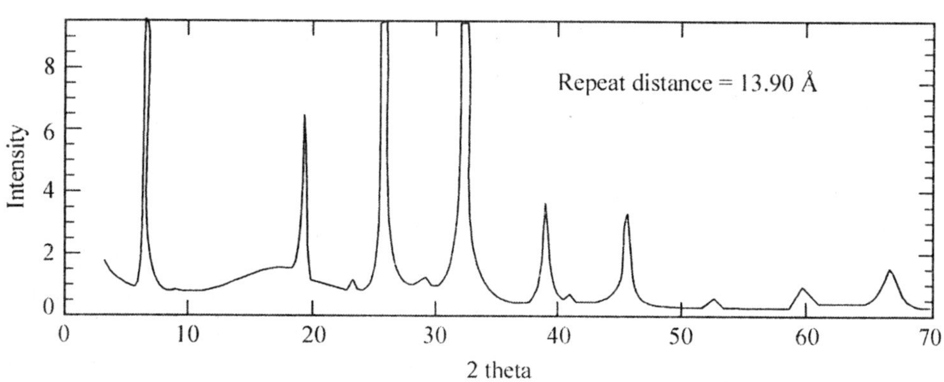

Fig. 55. X-ray diffraction pttern from a layered structure vermiculite clay.

Note 3:

A model calculation of unit cell size from diffraction data follows.

Example: The diffraction pattern of copper metal was measured with X-ray radiation of wavelength of 1.315 Å. The first order Bragg diffraction peak was found at an angle 2θ of 50.5°. Calculate the spacing between the diffracting planes in the copper metal.

The Bragg equation is

$$n \times \lambda = 2d \sin \theta$$

One can rearrange this equation for the unknown spacing d:

$$d = n \times \lambda/2 \sin \theta$$

θ is 25.25°, n = 1, and wavelength = 1.315 Å, and therefore

$$d = 1 \times 1.315/(2 \times 0.4266) = 1.541\text{Å}$$

Note 4. Instrumentation:

The X-ray diffraction experiment requires an X-ray source, the sample under investigation and a detector to pick up the diffracted X-rays. Fig. 56 is a schematic diagram of a powder X-ray diffractometer.

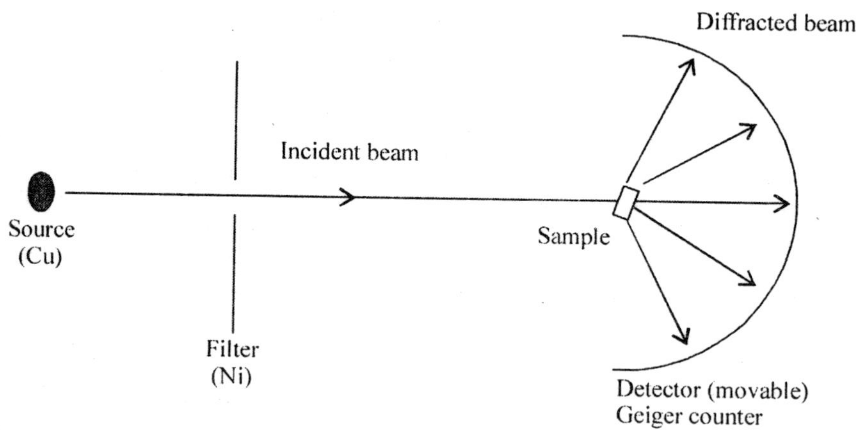

Fig. 56. Schematic of an X-ray powder diffractometer.

The X-ray radiation most commonly used is that emitted by copper, whose characteristic wavelength for the K radiation is = 1.5418 Å. When the incident beam strikes a powder sample, diffraction occurs in every possible orientation of 2θ. The diffracted beam may be detected by using a moveable detector such as a Geiger counter, which is connected to a chart recorder. In normal use, the counter is set to scan over a range of 2θ values at a constant angular velocity. Routinely, a 2θ range of 5 to 70° is sufficient to cover the most useful part of the powder pattern. The scanning speed of the counter is usually 2θ of 2 degrees min^{-1} and therefore, about 30 minutes are needed to obtain a trace.

EXPERIMENT 94
Surface Area Determination

Aim: To determine the surface area of alumina by nitrogen adsorption at 77 K.

Apparatus Required: The BET surface area determination equipment.

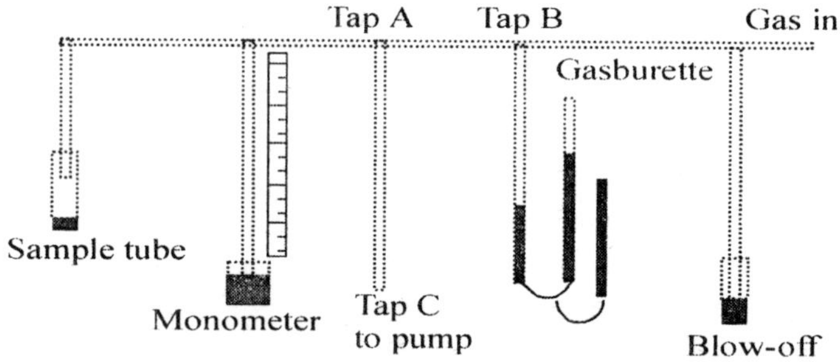

Fig. 57. Typical adsorption apparatus.

Chemicals Required: Any pure solid sample like iron oxide, etc.

Principle: Adsorption is the process by which atoms or molecules become attached to a surface. While adsorption may in principle occur at any surface, it has particular significance when a gas or liquid is in contact with a porous solid such as charcoal or alumina. Adsorption is described using an adsorption isotherm; this shows the amount of gas adsorbed as a function of pressure, at constant temperature. Various isotherms have been proposed based on different kinds of assumptions. The one well known isotherm called BET (Brunauer, Emmett and Teller) isotherm is more useful under these conditions. The BET result may be given as

$$[x/V\ (1 - x)] = (1/V_mC) + [(C - 1)\ x/V_mC] \qquad (1)$$

In equation (1), x is the relative pressure (p/p_0) at which a volume of gas V, measured at room temperature and pressure, is adsorbed. p is the pressure of the gas, and p_0 its saturation vapour pressure at the temperature of the reaction vessel containing the adsorbent. In this experiment nitrogen is the gas, and the adsorbent alumina is held at 77 K, the temperature of liquid nitrogen where, by definition, its vapour pressure p_o equals atmospheric pressure. V_m is the volume of adsorbed gas required to form a monolayer on the adsorbent, and is a constant for a given temperature. Equation (1) demonstrates that a plot of $x/[V(1 - x)]$ against x should give a straight line, whose intercept is $1/V_mC$ and slope is $(C - 1)/V_mC$. Thus from the slope and intercept, C and V_m can be found.

The constant C is given by:

$$C = [(a_1b_2)/(a_2b_1)] \exp [(E_1 - E_L)/RT] \quad (2)$$

where $(E_1 - E_L)$ is the difference between the average enthalpy of adsorption in the first layer and the enthalpy of liquefaction of the adsorbate. a_1, a_2, b_1 and b_2 are constants connected with the formation and evaporation of the first and higher layers of adsorbed molecules.

Brunauer and his co-workers have suggested that in certain circumstances (a_1b_2/a_2b_1) should be close to unity. If this is so, it would be possible to determine E_1 from a measurement of C from the isotherm.

The BET method yields a value for V_m, the volume of gas required to form a monolayer. Brunauer and Emmett suggested that the area of each adsorbed molecule could be found from the density of the liquid adsorbate according to the equation:

$$\text{Molecular area} = 4\ (0.866)\ [M/4\ (2)^{1/2}Nd]^{2/3}$$

where M is the molecular weight of the adsorbate, N is Avogadro's number and d the density of the liquefied adsorbent, which equals 0.808 g cm^{-3} for liquid nitrogen. Knowing V_m, the surface area can then be found.

Procedure: The gas line incorporates a mercury blow-off to prevent the pressure of nitrogen rising much above atmospheric pressure. Do not, however, rely upon this to prevent the pressure rising to a dangerous level: check that the pressure delivered by the regulator is never much above one atmosphere. The volume of gas in the apparatus is measured using a gas burette; pressure is measured using a mercury manometer. A sample tube containing alumina (aluminium oxide) can be heated in a small furnace. The procedure consists of the following steps.

Step 1. Removal of air from the apparatus:

Before starting the experiment, one should understand the operation of the vacuum line, the taps it contains and the nitrogen line.

Flush air from the gas burette as follows:

1. Check that the sample tube has been taken off the line.
2. Turn the three-way tap so that the nitrogen line is connected to the gas burette. Allow gas to bubble slowly and gently through the blow-off.
3. Lower the mercury reservoir to almost fill the burette with nitrogen (move the reservoir by holding the sliding clamp, not the reservoir itself); turn off the nitrogen supply.
4. Open the two-way tap (tap A) and turn the three-way tap (tap B) to connect the gas burette with the sample side of the vacuum line.
5. Raise the mercury reservoir carefully until the mercury level reaches the top of the graduated portion of the burette to flush the nitrogen to the atmosphere; take care that mercury does not enter the vacuum line.
6. Close the two-way tap and repeat this procedure twice more to remove all air from the burette.

Step 2. Determination of the dead volume:

When gas is allowed into the apparatus to adsorb on the alumina, some gas will remain unadsorbed in the left-hand section of the vacuum line. We must, therefore, determine the volume available to gas in this part of the line - the so-called "dead volume". Fill the gas burette with nitrogen. Turn the three-way tap so that the burette is connected only to the two-way tap, not to the nitrogen supply. Turn off the gas supply.

Clean the grease from the empty sample tube and from the socket on the gas line using a tissue and Volasil. If the sample tube appears dirty, wash it out with a little acetone and dry it using compressed air. Apply a thin ring of Apiezon T high temperature grease around the wide end of the tube and push it gently into position.

Switch on the pump and evacuate the manometer and sample tube until the pressure is constant. Close the two-way tap and monitor the pressure to check for leaks. Once you are sure the apparatus is gas-tight, adjust the mercury reservoir until the mercury levels in burette and levelling tube are equal. Record the manometer reading pi and read the initial volume of nitrogen V_i.

Open the two-way tap carefully to admit nitrogen from the burette into the sample space. When the pressure has increased by about 6 cm, as measured on the manometer, shut the two-way tap. Equalise the mercury levels in the burette and read the volume of nitrogen left in the burette (V_n); measure the corresponding pressure in the sample volume (p_n).

Admit further volumes of nitrogen to the apparatus and repeat the measurements to give a total of about 12 readings.

Step 3. Adsorption measurements:

Remove the sample tube and clean off the grease. Weigh accurately into the tube about 0.5 g alumina. Grease the joint (Apiezon T) and replace the tube. Pump out the apparatus. Place the furnace around the sample tube.

Place ceramic fibre around the top of the furnace and heat for 40 minutes with the pump running, to drive off adsorbed gases from the alumina. Allow the alumina to cool with the pump running.

Close the two-way tap, reverse the air leak on the conical flask to admit air to the rotary pump, then turn off the pump. Once the sample tube is cool enough to touch, place a small Dewar around it, and fill the Dewar completely with liquid nitrogen. You need now to admit small quantities (2-4 cm^3) of nitrogen to the vacuum line.

Fill the burette with nitrogen, then very cautiously turn the three-way tap so that the burette is connected by a leak of gas to the sample region.

Admit 2-4 cm^3 of nitrogen (as measured by the gas burette) to the sample tube, using the procedure described in the previous section. (Don't forget that a movement of 4 cm by the mercury up the gas burette will amount to a volume of more than 4 cm^3 once you equalise the mercury levels). When between 2 and 4 cm^3 nitrogen has been admitted, close the two-way tap, equalise the mercury levels and determine the volume of gas admitted, measured at room pressure. Tap the manometer gently to prevent the mercury from sticking. Read and record the pressure above the sample every 30 seconds until the readings stabilise (stabilisation should require between 3 and 6 minutes, with the longer times required later in the experiment).

Record the equilibrium pressure p. Now continue adding nitrogen in smaller steps, (about 2 to 3 cm^3 each time) until a pressure of about 2/3 atmosphere has been reached inside the apparatus.

On completion of the experiment remove the Dewar, release the pressure in the line and remove the sample tube. If you do not do this, pressure will rise in the apparatus and may cause it to burst. Turn off the nitrogen cylinder. Measure and record room temperature and pressure.

Treatment of Data:

Tabulate your gas pressure and volume readings. Be sure that the pressure you quote is the actual pressure in the apparatus, not simply the reading of the manometer.

Calculate the dead volume of the apparatus, as follows: Let the volume of sample tube and manometer combined be V'. The volume of gas admitted to this part of the apparatus is $(V_i - V_n)$ at the ambient pressure, p_{atm}. When the gas enters the volume V', it creates a pressure p_n, and therefore:

$$P_n V' = p_{atm} (V_i - V_n)$$

Plot a graph of measured pressure against the volume of gas admitted and thus determine the dead volume (the volume of gas which would need to be admitted to give one atmosphere pressure within the apparatus).

Surface area calculation:

For the second part of the experiment, tabulate for each addition of gas: the manometer reading, the actual pressure in the gas line, the burette reading, and the volume of gas, at room T and P, admitted to the line (remember your readings are cumulative). Use the graph you constructed above to determine the volume of gas not adsorbed. (You should be able to do this readily, since your graph relates the pressure of unadsorbed gas in the apparatus to the volume of gas needed to create such a pressure). Hence, for each reading determine the volume of gas adsorbed, measured at room temperature T and P.

Plot the volume of gas adsorbed as a function of pressure of gas above the sample.

Test the validity of the BET isotherm for this system by using equation (1) to make a suitable plot. Fit a straight line through your points, determine V_m and thus find the surface area of 1 g alumina.

Estimate E_1, the enthalpy of adsorption of nitrogen on alumina, given that E_L is 5.6 kJ mol^{-1}.

An expression for the time an adsorbed molecule stays on the surface before evaporating is $\tau = \tau_0 \exp E_1/RT$, where τ_0 is the time for a single vibration of the molecule in the absorbed state (which one can assume to be 10^{-13} s).

Result:

Calculate and report the time an average nitrogen molecule spends on the surface.

Comment on your results.

Compare your value for E_1 with that for hydrogen adsorbed on nickel (125 kJ mol^{-1}).

Report the surface area of the sample of Alumina.

EXPERIMENT 95

Excited State Propeperties of Molecules

Aim: To determine the acidity constants of the ground state and (lowest electronically) excited states of 2-naphthol [denoted as (ArOH)] in aqueous solution.

Apparatus Required: Spectrophotometer, Fluorimeter, pH meter and graduated pipettes (5 mL).

Chemicals Required: 2-Naphthol, NaOH, HCl and NH_4Cl / NH_4OH buffers.

Principle: In aqueous solution, 2-naphthol behaves as a weak acid forming the hydronium ion and its conjugate base the naphthoxy ion, ArO^-

$$ArOH + H_2O \longleftrightarrow ArO^- + H_3O^+$$

The relative energies of the free acid (ArOH) and its conjugate base (the naphthoxy ion ArO^-) are shown for both the electronic ground state (S_0) and

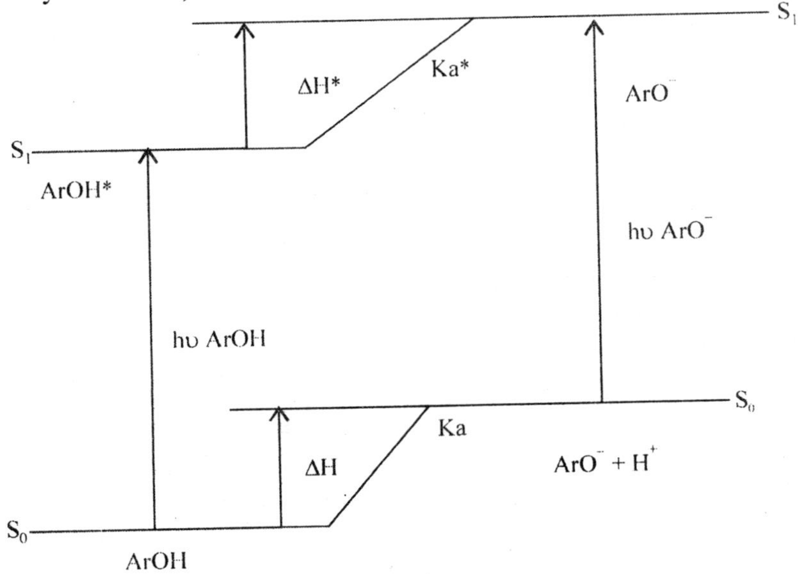

Fig. 58. Schematic diagram of the ground and first excited single state energies of free 2-naphthol and its conjugate base naphthoxy ion in aqueous solution.

(lowest) excited state (S_1) in aquesous solution. Each of the anions are elevated with respect to its free acid by an energy, ΔH and ΔH^*, respectively. These can be recognized as the enthalpies of deprotonation of the ground state and excited molecule of 2-naphthol. Both the ground state 2-naphthol and the naphthoxy anion can be transformed to the respective excited states by absorption of photons of energy; equal to $h\upsilon_{ArOH}$ and $h\upsilon_{ArO^-}$. For simplicity we have not considered the vibrational levels involved in the transitions.

The free energy change for the two states can be given as

$$\Delta G = \Delta H - T\,\Delta S = -RT \ln K_a$$

and

$$\Delta G^* = \Delta H^* - T\,\Delta S^* = -RT \ln K^*_a.$$

for the S_0 and S_1 states respectively. If one were to assume that the entropies of dissociation of ArOH and ArO^- are almost the same, then it follows that

$$\Delta H - \Delta H^* = -RT \ln (K_a^*/K_a)$$

and from the above figure one can deduce that

$$\Delta H + N_A h\upsilon_{ArO^-} - \Delta H^* = N_A h\upsilon_{ArOH}$$

where h is the Planck's constant and N_A is the Avogadro's number for the conversion to molar basis.

Therefore, one can write

$$\ln (K_a^*/K_a) = [N_A h\{\upsilon_{ArO^-} - \upsilon_{ArOH}\}]/RT.$$

This can be rewritten as

$$pK_a^* = pK_a - [N_A h c/2.3 RT]\{\upsilon_{ArOH} - \upsilon A_{rO^-}\}.$$

Procedure: Record the absorption spectra of two solutions of ArOH (approximately 10^{-4} M but the actual concentration being known) one at low pH and other at high pH (using concentrated stock solutions of HCl and NaOH approximately to create $[H^+]$ and$[OH^-]$ is of the order of 0.02 M. Now obtain absorption spectra of ArOH at various intermediate values of pH by using buffer solutions of NH_4OH/NH_4Cl keeping the 2 –naphthol concentration (known) the same in each of these cases. Measure the pH of each of the solutions after recording the spectra. Since the bulk 2–naphthol concentration that is $[ArOH]^+$ $[ArO^-]$ is invariant, the spectra when properly overlapped should intersect at a common wavelength called the isostestic point (equal absorption). Obtain the fluorescence spectra of naphthol and its conjugate base. Record the fluorescence spectra at intermediate pH solutions.

Treatment of Data: Since the $[ArOH]_0$ is the same in all the solutions the following material balance holds good in all the solutions.

$$[ArOH]_0 = [ArOH] + [ArO^-].$$

According to Beer's law one can write

$$A\ (\lambda_{max}\ ArOH) = \varepsilon_{ArOH}\ [ArOH] + \varepsilon A_{rO^-}\ [ArO^-]$$

where A is the absorbance (for a 1 cm path length) and ε_{ArOH} and ε_{ArO^-} are the molar absorptivity coefficients of the free acid and conjugate base at λ_{max}(ArOH) respectively.

Combining these two equations one obtains

$$A(\lambda_{max} ArOH) = \{\varepsilon_{ArOH} - åA_{rO^-}\}\ [ArOH] + \{\varepsilon A_{rO^-}\}\ [ArOH]_o$$

This relation allows the determination of [ArOH] under equilibrium conditions. Once this is known then the $[ArO^-]$ can be calculated and the acidity constant in the ground state can be calculated as

$$K_a = [H_3O^+]\ [ArO^-]/[ArOH]$$

Calculate the value of pK_a^* from the value of pK_a and also from the value of $(\upsilon_{ArOH} - \upsilon A_{rO^-})$ determined from the fluorescence maxima of the two species namely the free acid and conjugate base.

Result:

Report the pK_a values of the ground state and first excited state of 2-naphthol. Comment on the relative acidity of the two molecules.

Note:

This experiment can be repeated for other molecules like 2-naphthoic acid, acridine, and also quinoline.

EXPERIMENT 96
UV—Visible Spectroscopy

Aim: To study the effect of extended conjugation on the wave length of maximum absorption (λ_{max}) of organic compounds.

Apparatus Required: UV–Visible spectrophotometer, cuvettes, graduated pipettes and volumetric flasks.

Chemicals Required: p-Nitrophenol, ethanol and NaOH.

Principle: Ultraviolet and visible light, when incident on an organic molecule having π electrons and/or non bonded electrons would excite them and results in absorption of the light. When the energy difference between the highest occupied molecular orbital (HOMO) and the lowest unoccupied molecular orbital (LUMO) of the molecule exactly matches the energy of the incident light , maximum absorption occurs. The wave length at which this maximum absorption occurs is referred to as the λ_{max}. The position of this λ_{max} is highly influenced by the extent of conjugation in the molecule. Generally, the higher the conjugation the higher the value for $\lambda_{max.}$

When a base is added to p-nitrophenol (λ_{max} = 320 nm) it is ionized to p-nitrophenoxide (λ_{max} = 400 nm) in which extended conjugation is possible as shown below. Hence, addition of the base to p-nitrophenol results in a bathochromic shift.

Procedure: Prepare a 0.001 M solution of p-nitrophenol in a solvent consisting of 20 % alcohol and 80 % water. Record its spectrum by scanning from 270 nm to 450 nm. Then add a small bit of NaOH crystal , say 0.05 g, to this solution and dissolve it completely. The solution may turn slightly yellow. Record the spectrum for this solution also in the same wave length range, preferably on the same sheet.

Result:

Report the λ_{max} obtained before and after addition of NaOH.

Comment on your observation.

Note:

The importance of conjugation: A comparison of the absorption spectrum of 1-pentene, λ_{max} = 178 nm, with that of isoprene demonstrates the importance of chromophore conjugation. The spectrum given in (Fig. 59) shows that conjugation of double and triple bonds shifts the absorption maximum to longer wavelengths. From the polyene spectra shown in (Fig.59), it is clear that each additional double bond in the conjugated pi-electron system shifts the absorption maximum by about 30 nm in the same direction.

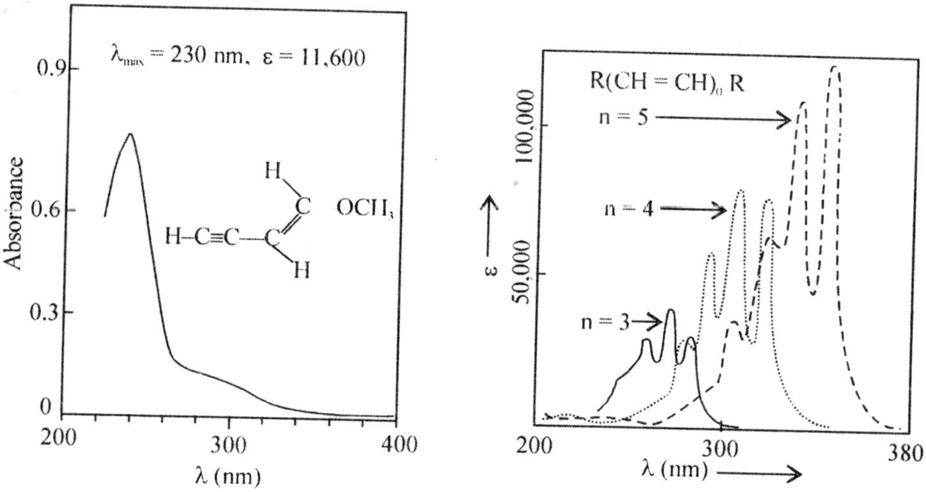

Fig. 59. Typical electronic spectra showing the effect of conjugation and the absorption maximum.

The reader may practice answering questions of the following types to make oneself familiar with the application of UV-Visible spectroscopy.

Typical questions are:

1. Comment on the absorption maximum of $C_6H_5NH_2$ and $C_6H_5NH_3^+ Cl^-$ and account your answer.
2. Consider the following two isomers $H_2C = C(CH_3) - CH_2COCH_3$ and $CH_3 - CH = C(CH_3) - COCH_3$ One of them absorbs at 235 nm with high ε while the other has no high intensity absorption above 220 mm. Which one has a high absorption at 235 nm and why?

3. Why one does not observe n > λ^* transition in the spectrum of trimethyl amine in acidic solution?
4. The spectrum of methyl chloride shows an absorption band at 173 nm while methyl iodide shows absorption band at 259nm. Account for this observation.
5. The reader is advised to select many other problems and attempt to solve them.

EXPERIMENT 97
Infrared Spectroscopy

Aim: To find out the frequency of the normal vibrational modes of carbon dioxide in the infra red region.

Apparatus Required: Infra red spectrometer, vacuum system, a gas cell.

Chemicals Required: Pure carbon dioxide gas.

Principle: Consider a linear, triatomic molecule say carbon dioxide. From the statistical mechanical treatment of gaseous molecular heat capacities, one can obtain the number of degrees of freedom. For a linear triatomic molecule, $3N$ 5 vibrational modes are possible, where N is the number of atoms in the molecule. As there are three atoms in CO_2 it will have 4 vibrational degrees of freedom remain. These 4 degrees of freedom are known as normal modes of vibration and are conventionally labeled as follows:

υ_1 = symmetric stretch,
υ_2 and υ_2' = bending,
υ_3 = asymmetric stretch.

In order for a characteristic vibration to be infrared active, there must be a change in the dipole moment of the molecule as it executes the vibrational motion. Each IR-active vibration will appear as a band of lines due to simultaneous changes in the rotational energy. However, for many molecules the rotational structure will not be resolvable.

The base frequency for the symmetric stretch mode is $\upsilon_1 = 1288\ cm^{-1}$ and for the bending mode $\upsilon_2 = 667 cm^{-1}$. For the asymmetric stretch mode, the base frequency is much higher at $\upsilon_3 = 2349 cm^{-1}$.

Procedure:

1. Fill an IR gas cell with one atmosphere of CO_2. Either use the portable vacuum line or flush the cell with CO_2 from one of the larger tanks.

2. Scan the whole range (4000 – 600 cm^{-1}).
3. Identify the strong peaks; these are the normal modes.
4. With the aid of the vacuum system, reduce the pressure of CO_2 in the cell until the resolution of the peaks is maximized.
5. The (perpendicular) bending mode has a central peak, called the Q-branch. The (parallel) stretching modes have only P and R branches. Assign the major peaks, i.e., υ_2 and υ_3.
6. "Combination bands" and "overtones" sometimes appear, although weaker than single normal modes. Look at the 1 atmosphere scan. Identify and label peaks corresponding to the following: $\upsilon_1 + \upsilon_2$, $\upsilon_1 + \upsilon_3$, $3\upsilon_2 + \upsilon_3$, $3\upsilon_2$, $2\upsilon_1 + \upsilon_3$. With the available information, calculate the frequency of the IR-inactive mode.
7. Organize the observed transition energies into a table and label them with the appropriate linear combination of normal modes. Include inactive modes and anharmonicities (deviation from the harmonic oscillator model). Discuss the contents of your table and other relevant issues pertaining to the experiment.

Result:

Report the frequencies of the normal modes of vibration and the overtones for carbon dioxide.

Note:

Read detailed principles of IR spectroscopy from standard spectroscopy text books and attempt the following exercises to sharpen your understanding.

Typical exercises are:

1. With the help of suitable diagram show the normal modes of vibration in water. Comment on which ones are infrared active and also which are degenerate.
2. Calculate the number of normal vibration modes possible for ethane and benzene.
3. The fundamental vibration frequency of carbon monoxide molecule is 2143 cm^{-1}. Calculate the force constant of the C – O bond.

   [1858 N m^{-1}]
4. The force constant for HF molecule is 970 N m^{-1}. Calculate the frequency of radiation to excite the molecules from v = 0 to v = 1 state.

   [4158.9 cm^{-1}]

5. The force constants of HF and DF are the same. Find the zero point energy of these two molecules, the fundamental vibrational frequencies and the difference between the zero point energies of HF and DF.
6. Tabulate the values of force constants for a set of typical functional groups like N – N, N = N, N ≡ N and similar other species and comment on the relative magnitudes.
7. The vibrational constants of one of the excited states of bromine molecule is 151.2 cm^{-1}, and $W_e x_e$ = 0.33 cm^{-1}. What is the dissociation energy of bromine molecule in this state? [207 kJ mol^{-1}]
8. Pictorially show the PQR branches of a vibrational spectrum of a harmonic oscillator.
9. The absorption of infrared radiation by $^{12}C^{16}O$ molecule shows a band at 2135 cm^{-1}.
   (i) Calculate the corresponding wavelength in CM^{-1} and the frequency in Hertz.
   (ii) Calculate the value of force constant in Nm^{-1}.
   (iii) What will the frequency if the ^{16}O is replaced by ^{18}O? What will the value of force constant for $^{12}C^{18}O$?
   (iv) Calculate the zero point energy of $^{12}C^{16}O$ in kJ mol^{-1}. [4.684 × 10^{-4} cm, 6.405 × 10^{13} hz, 1.843 × 10^{3} Nm^{-1}, 2083 cm^{-1}, 12.78 kJ mol^{-1}.] The reference spectrum can be seen in any standard book, one can see G.M. Barrow, The structure of Molecules, W. A. Benjamin (1964) p.72.
10. The vibration rotation spectrum of HBr can be seen from any standard text book. One source is G.M. Barrow, The structure of Molecules, W. A.Benjamin (1964) 93.
    (i) Calculate the frequency of the centre of the band for HBr. Identify the P and R branches. Why is the Q branch absent?
    (ii) Calculate the spacing of the lines and the bond length of HBr.
    (iii) Calculate the position of the centre of the band for DBr.
    (iv) What will be the numerical magnitude of the centre of the band and the spacing of the lines for HCl? [2555 cm^{-1}Q branch is absent because the transition with $\Delta J = 0$ are not allowed for diatomic molecules (NO is exception); 16.90 cm^{-1};1.42 Å; 1818 cm^{-1}; the spacing of the rotational lines of HCl is higher than that of the rotational lines of HBr; reason out this statement.]

EXPERIMENT 98
Electron Spin Resonance Spectroscopy

Aim: To understand the basic coupling laws which determine the appearance of Electron Spin Resonance (ESR) Spectra. Also , from an ESR spectra how to determine the electron densities around the ring of the aromatic radical. (These electron densities can also be calculated using simple Huckel molecular orbital theory (HMO). This comparison will help to understand some basis of MO theory also.)

Apparatus Required: ESR spectrometer, Erlenmeyer flasks, pipettes.

Chemicals Required: Anthracene, any odd electron species (usually 2, 2, 6, 6-tetramethyl-1-piperidinyloxy (TEMPO) radical is used).

Principle: ESR is exhibited only by molecules which contain unpaired electrons. Therefore, we will consider only free radicals. The electron has two possible observable values of its spin angular momentum $m_s h/2\pi$, where $m_s = \pm \frac{1}{2}$; each of these yields a different value of its magnetic moment, defined as $\mu = \pm 2g\mu B$. Here μB is the basic unit of electron magnetism, the Bohr magneton ($he/2\pi m = 9.2737 \times 10^{-24}$ J Tesla^{-1}) and g is a constant. = 2.0023 for a free electron. In radicals spin-orbit coupling changes the total angular momentum of the electron and hence the magnetic moment and the g-value. The precise value of g is therefore characteristic of the specific radical involved.

When a molecule with a magnetic moment is placed in a magnetic field B, it acquires an energy equal to $-\mu B$, and hence the external field removes the degeneracy of the electron spin states to give two energy levels:

$$E_{1/2} = -1/2\ g\ \mu_B B \text{ and } E_{-1/2} = +1/2\ g\ \mu_B B$$

The populations of electrons in these two levels is governed by normal Boltzmann distribution at thermal equilibrium.

The ESR experiment consists simply in inducing transitions between the two levels with radiation of frequency ν such that

$$h\upsilon = E_{-1/2} - E_{1/2} = g\,\mu_B B$$

A magnetic field of 0.33 Tesla (3300 gauss) is normally used in which transitions occur at a frequency of about 9.5 GHz; the spectrum is displayed by sweeping the field whilst keeping the frequency constant.

The reader is advised to read the basis of ESR spectrometer, the components that go to make a spectrometer and the method of operation including the phase sensitive detection used to increase signal to noise ratio from any standard text book.

Procedure:

Hyperfine structure illustrated by coupling in a nitroxide radical:

As stated above, all ESR spectra should consist of a single transition centred at a field position which depends upon the g-value of the radical. However this will be observed only when the species contain nuclei with zero spin. In the case of molecules or species containing nuclei possessing spins, additional lines are observed arising from the coupling of the electron and nuclear spins. These additional lines are termed as hyperfine structure with its own characteristic coupling constant. For demonstration purposes let us consider 2, 2, 6, 6-tetramethyl-1-piperidinyloxy (TEMPO) radical, which is widely used as a 'spin label' in biological studies. Dissolve about a dozen crystals of TEMPO in a few mL of dry 1, 2-dimethoxyethane, then pipette some solution into an EPR tube so that the tube is half full. Cap the tube and record the ESR spectrum. Adjust the measurement conditions, and the concentration of the solution, so as to obtain the best possible separation between the lines in the spectrum.

Hyperfine structure illustrated by coupling to equivalent H-nuclei: The benzoquinone radical anion:

One can prepare a sample of the radical anion of p-benzoquinone by reduction with hydroxyl ions. The radical is stabilized by the presence of 1, 2 – dimethoxyethane. Prepare 10 mL of a 50/50 mixture of dry propan-2-ol and dry 1,2-dimethoxyethane. Add a single pellet of potassium hydroxide and stir. Deoxygenate the solution by passing a gentle stream of nitrogen. Add about half a microspatula full of benzoquinone; a green-black solution will develop immediately. Mix gently, then use a Pasteur pipette to introduce the solution into a clean, dry EPR tube. Record the spectrum of the benzoquinone radical anion which should give a simple 5-line pattern.

Conclusion:

A radical in which an electron couples to n spin ½ (H) nuclei exhibits a spectrum of (n + 1) lines. These lines are equally spaced and have relative intensities given by the coefficients of the terms in the expansion of $(1 + x)^n$. A radical in which the electron couples to a nucleus of spin I has a spectrum split into (2I + 1) equally-spaced lines of equal intensity. In each case the hyperfine coupling constant is measured directly from the observed splittings between the lines.

The origin of the hyperfine coupling is the 'Fermi contact interaction' and the energy of the interaction is given by A.I.S where I is the nuclear spin vector and S is the electron spin vector. A is the hyperfine coupling constant and depends upon the electron density at the nucleus, which is proportional to the square of the electron wave function at the nucleus $|\Psi(0)|^2$. Since only electrons in 's' orbitals have any probability of being at the nucleus, it is expected that only electrons in orbitals which possess some 's' character will exhibit hyperfine coupling.

The coupling rules stem directly from a consideration of the number of ways in which nuclear spins can be arranged. I can take (2I + 1) values of m_I, the observable component of I, in the presence of a field and there are consequently (2I + 1) different values of the interaction energy. The total energy of the electron consists of a combination of these values with the energy it possesses from being in an external field. With selection rules $\Delta m_S = \pm 1$ and $\Delta m_I = 0$, this implies that a radical in which an electron couples to a single proton will have a 2-line spectrum. Similarly, a spin 1 ^{14}N nucleus will lead to a 3-line spectrum.

When equivalent nuclei are present, the value of I used must be the vector combination of the individual spins: for 2 equivalent spin ½ nuclei these are {+½ +½}, {+½ –½}, {–½ +½} and {–½ –½}, giving resultant values of +1, 0, –1 with statistical weights of 1, 2 and 1 respectively. The ESR spectrum is consequently a 1 : 2 : 1 triplet. If coupling occurs to two non-equivalent spin ½ nuclei each splits the ESR line independently in turn: the 'first' yields a doublet and the 'second' then splits each component of the doublet into a further doublet of different splitting. The spectrum consists of a quartet inside which each coupling constant can be measured twice.

Indirect coupling and the McConnell equation:

In aromatic radicals, the odd electron occupies a molecular orbital which is delocalised over the atomic framework of the radical. Being in a π orbital, it has zero probability of being in the plane of the ring where the H atoms lie - the π orbital has a node there. However, despite our conclusion above on 's' character, hyperfine interactions with the H-atoms are observed. We conclude that a mechanism exists by which the 1s hydrogen orbitals acquire some unpaired spin density.

This occurs via a 'spin polarization' mechanism and originates in the exchange interaction between the σ electrons in the C-H bond and thc π electrons in the ring. Consider a single C-H fragment. We can draw the electronic structure in two ways, which differ in the electron spin orientations in the sp^2 C and s H orbitals.

In terms of the normal 'pairing' approximation each arrangement would be equally likely. However the exchange interaction causes the arrangement in which the electron spins in the two carbon atoms are parallel to be slightly preferred (according to Hund's rule electrons in degenerate orbitals have parallel spins). As a result the electrons in the C-H bond are polarized slightly so that one electron spin is favoured over the other, leading to a corresponding predominance of one spin state of the electron in the hydrogen s orbital over the other. This provides the unpaired spin density at the nucleus required for coupling.

In aromatic radicals delocalisation causes the average unpaired spin population of a particular carbon $2p_z$ orbital to be < 1, and the amount of spin polarization it causes in the C-H bond depends upon its precise value. This is termed the 'π-electron spin density, ρ'. The McConnell equation states that the hyperfine coupling constant A is proportional to ρ:

$$A = Q \cdot \rho$$

where Q is a constant for the type of atom with the unpaired spin density. This important relationship allows a direct link between experiment and theory, for ρ can often be calculated by Huckel MO methods. The constant Q can be determined experimentally and for carbon atoms this is performed conveniently by observing the ESR spectrum of the benzene radical anion,$C_6H_6^-$. This spectrum has seven lines, equally spaced and with intensity ratios showing that the electron couples to six equivalent protons. This confirms that it occupies a molecular orbital delocalised over the ring-a nice demonstration of basic orbital theory. Symmetry requires all the carbon atoms to be equivalent so that the observed splitting of 3.75 gauss (0.375 mT) represents the value of A when $\rho = 1/6$.

Evaluate Q_C; it will be needed to obtain ρ_c from the observed splittings in the anthracene cation.

The anthracene cation:

The spectrum of anthracene is more difficult to obtain than those of TEMPO and benzoquinone. In order to record a good spectrum, it is important that you follow these instructions carefully.

Check the settings on the ESR control unit of the phase, scan range and centre field.

These settings will provide a useful guide to where to start looking for the anthracene resonances. Typical – though not necessarily optimum – values are:

Centre field 530	Range 455	Speed 390
100 Hz Mod middle setting	Attenuator 0	Phase 160

Scan time 60 seconds.

Prepare a solution of anthracene of concentration approximately 0.02 M as follows. First put on safety glasses! Now add about a microspatula-full of anthracene to a clean 100 mL Erlenmeyer flask - there should be perhaps 50-100 tiny crystals in total. Carefully add about 25ml concentrated sulphuric acid, and mix by swirling for thirty seconds. Be careful to avoid spills. Try to maintain a smooth liquid surface, since the more you break up the liquid into droplets the more you will introduce oxygen and water vapour into the solution, both of which will reduce the concentration of free radical.

The solution should be green and then quite rapidly turn very dark green or black. If the colour remains light the solution is much too dilute - you need a solution in which there is extensive charring (production of carbon, giving the black colour) to have a chance of seeing the spectrum. Transfer a portion of the sample to a clean and dry EPR tube. Check that no acid whatsoever remains on the outside of the tube, then place in the sample area between the magnet poles.

The sample is difficult to tune, (because of the high dielectric constant of sulphuric acid) so rapidly carry the normal tuning procedure as far as possible, then attempt to find the spectrum. If the spectrum -once found-shows excessive noise, you may tune the cavity while a spectrum is being gathered.

The radicals should be detectable for no more than 5 minutes, so you are unlikely to obtain a good spectrum from your first sample. Once you have found where the spectrum lies, immediately save a spectrum, then concentrate on zeroing in on the lines, adjusting the scan range, centre field and scan time so that the spectrum is as well defined as possible. The quality of the spectrum is particularly dependent upon the speed at which the field is scanned and the setting of the 100 Hz field. A well-resolved spectrum on this instrument shows 21 lines. (This is far fewer than the theoretical number-construction of the stick diagram latter will help you understand why this is.)

When you have determined suitable operating parameters, prepare a fresh sample if necessary and record a final spectrum.

Calculation:

Use the McConnell equation and the values for ρ_c to calculate the coupling constants A_H. By drawing a diagram in which the effects of the different couplings are shown as causing successive splittings of a single original line, deduce the underlying coupling pattern for anthracene, and compare with your experimental spectrum.

Result:

Report the coupling pattern of the species studied.

Note:

The following articles are suggested for background reading.

1. Magnetic resonance, K.A.McLauchlan, Clarendon Press. Introduction to magnetic resonance. A.Carrington and A.D.McLachlan, Chapman & Hall.
2. Electron Spin Resonance of aromatic radicals and radical ions. Quarterly reviews of the Chemical Society, 1968, 17, 67. Electron paramagnetic resonance: elementary theory and practical applications. J.A.Weil, J.R.Bolton and J.E.Wertz, Wiley.

EXPERIMENT 99
Nuclear Magnetic Resonance Spectroscopy

Aim: To record the proton nuclear magnetic resonance spectrum of ethanol and understand the splitting pattern.

Apparatus Required: Proton NMR spectrometer and accessories.

Chemicals Required: Pure ethanol, sulphuric acid and tetramethylsilane.

Principle: When a nucleus having a resultant spin as in the case of a proton is placed in an effective magnetic field its energy levels split into two with their energy difference $\Delta E = [\gamma\ h\ B/2\pi]$. The nucleus is spinning on its axis. In the presence of a magnetic field, this axis of rotation will precess around the magnetic field; When simultaneously a radio frequency is also applied to the system, it will be absorbed more when the applied RF is in resonance with the precessional frequency. The intensity of this absorption can be traced by suitable detectors. The position of these absorption peaks depend on the chemical environment of the proton(defined in terms of the chemical shift) and the nature of the peaks depend on the actual neighbours (defined in terms of the spin-spin coupling constants).

Procedure: Record the proton nmr spectra for pure ethanol using the tetramethysilane as the reference compound. Again record the spectra for an ethanol sample to which a small amount of dilute sulphuric acid is added. Note down the positions of the peaks in terms of chemical shift (δ) and observe the splitting pattern of the peaks.

Result:

Report the positions of the methyl, methylene and the hydroxyl protons observed in both the spectra. Comment on your observations.

Note 1:

The proton NMR spectrum of ethanol (a) in the presence of traces of acidic impurities and (b) in a highly pure state are given in Organic chemistry, W.Kemp.

Note 2:

In NMR spectroscopy, one is performing experiments on the nuclei of atoms, and not on the electrons. The chemical environment of specific nuclei is deduced from the information obtained around a given nucleus.

Note 3. Nuclear spin and the splitting of energy levels in a magnetic field:

Subatomic particles (electrons, protons and neutrons) can be imagined as spinning on their axes. In many atoms (such as ^{12}C) these spins are paired against each other, such that the nucleus of the atom has no overall spin. However, in some atoms (such as ^{1}H and ^{13}C) the nucleus does possess an overall net spin. The rules for determining the net spin of a nucleus are as follows:

1. If the number of neutrons and the number of protons are both even, then the nucleus has no net spin.
2. If the number of neutrons plus the number of protons is odd, then the nucleus has a half-integer spin (i.e. 1/2, 3/2, 5/2).
3. If the number of neutrons and the number of protons are both odd, then the nucleus has an integer spin (i.e. 1, 2, 3) The overall spin, I, is important parameter. A nucleus of spin I will have 2I + 1 possible orientations. A nucleus with spin 1/2 will have 2 possible orientations. In the absence of an external magnetic field, these orientations are of equal energy. In the presence of an applied magnetic field, the energy levels are split with each level is associated with a given magnetic quantum number, m.

Note:

When the nucleus is in a magnetic field, the initial populations of the energy levels are governed by the Boltzmann distribution, and the transition energy can be calculated. The nucleus has a positive charge and is spinning. This generates a small magnetic field. The nucleus therefore possesses a magnetic moment, m, which is proportional to its spin, I,

$$\mu = (\gamma I h/2\pi)$$

where the constant, γ, is called the magnetogyric ratio and is a fundamental nuclear constant which has a different value for every nucleus, h is Planck's constant. The energy of a particular energy level is given by;

$$E = -[(\gamma h)/2\pi]\ m.B$$

where B is the strength of the magnetic field at the nucleus. The difference in energy between levels (the transition energy) can be found from $\Delta E = [\gamma\ h\ B/2\pi]$.This means that the transition energy is a function of the applied magnetic field, B. The transition energy also depends on the magnitude of the magnetogyric ratio, the larger it is the larger is the ΔE.
The absorption of radiation by a nucleus in a magnetic field: Imagine a nucleus (of spin 1/2) is in a magnetic field. This nucleus is in the lower energy level (i.e. its magnetic moment does not oppose the applied field). The nucleus is spinning on its axis. In the presence of a magnetic field, this axis of rotation will precess around the magnetic field.

Note 5. Definitions:

The frequency of precession is termed the *Larmor frequency*, which is identical to the transition frequency. The potential energy of the precessing nucleus is given by;

$$E = -\mu\ B \cos\theta,$$

where θ is the angle between the direction of the applied field and the axis of nuclear rotation. If energy is absorbed by the nucleus, then the angle of precession, θ, will change. For a nucleus of spin 1/2, absorption of radiation "flips" the magnetic moment so that it opposes the applied field (the higher energy state).

Relaxation processes: Let us now consider how the nuclei in high energy state relax and come to the ground state, that is we are examining the possible relaxation processes for the energy states of the nuclei. Emission of radiation is insignificant because the probability of re-emission of photons varies with the cube of the frequency. At radio frequencies, re-emission is negligible. One has to consider only non-radiative relaxation processes.

There are two major relaxation processes; (i) Spin-lattice (longitudinal) relaxation and (ii) Spin-spin (transverse) relaxation.

Spin-lattice relaxation : NMR active nuclei is contained in a lattice in the given sample. Nuclei in the lattice have vibrational and rotational motions thus creating a complex magnetic field which is called the lattice field with many components. Some of these components can be equal in frequency and phase to the Larmor frequency of the nuclei of interest. These components of the lattice field can interact with nuclei in the higher energy state, and cause them to lose energy (returning to the lower state). The energy that a nucleus loses increases the amount of vibration and rotation within the lattice causing a slight increase in the temperature of the sample.

The relaxation time, T_1 (the average lifetime of nuclei in the higher energy state) is dependant on the magnetogyric ratio of the nucleus and the mobility of the lattice. As mobility increases, the vibrational and rotational frequencies increase, making it more likely for a component of the lattice field to be able to interact with excited nuclei. However, at extremely high mobilities, the probability of a component of the lattice field being able to interact with excited nuclei decreases.

Spin-spin relaxation: Spin-spin relaxation deals with the interaction between neighbouring nuclei with identical precessional frequencies but differing magnetic quantum states. In this situation, the nuclei can exchange quantum states; a nucleus in the lower energy level will be excited, while the excited nucleus relaxes to the lower energy state. There is no net change in the populations of the energy states, but the average lifetime of a nucleus in the excited state will decrease. This can result in line-broadening.

Chemical shift: The magnetic field at the nucleus is not equal to the applied magnetic field; electrons around the nucleus shield it from the applied field. The difference between the applied magnetic field and the field at the nucleus is termed the nuclear shielding.

Consider the s-electrons in a molecule. They have spherical symmetry and circulate in the applied field, producing a magnetic field which opposes the applied field. This means that the applied field strength must be increased for the nucleus to absorb at its transition frequency. This upfield shift is also termed diamagnetic shift. Electrons in p-orbitals have no spherical symmetry. They produce comparatively large magnetic fields at the nucleus, which give a *low field shift*. This "deshielding" is termed paramagnetic shift. In proton (^1H) NMR, p-orbitals play no part, and hence only a small chemical shift (10 ppm) is usually observed. Chemical shift is a function of the nucleus and its environment. It is measured relative to a reference compound. For ^1H NMR, the reference is usually tetramethylsilane, $Si\ (CH_3)_4$.

Spin-spin coupling: Consider the structure of ethanol;

$$CH_3 - CH_2 - OH$$

The ^1H NMR spectrum of ethanol shows that the methyl peak has been split into three peaks *(a triplet)* and the methylene peak has been split into four peaks (*a quartet*). This occurs because there is a small interaction (coupling) between the two groups of protons. The spacings between the peaks of the methyl triplet are equal to the spacings between the peaks of the methylene quartet. This spacing is measured in Hertz and is called the coupling constant, J.

Why the methyl peak is split into a triplet? Consider the methylene protons. There are two of them, and each can have one of two possible orientations (aligned with or opposed to the applied field). This gives a total of four possible states.

In the first possible combination, spins are paired and opposed to the field. This has the effect of reducing the field experienced by the methyl protons; therefore a slightly higher field is needed to bring them to resonance, resulting in an upfield shift. Neither combination of spins opposed to each other has an effect on the methyl peak. The spins paired in the direction of the field produce a downfield shift. Hence, the methyl peak is split into three, with the ratio of areas 1 : 2 : 1.

Similarly, the effect of the methyl protons on the methylene protons is such that there are eight possible spin combinations for the three methyl protons.

Out of these eight groups, there are two groups of three magnetically equivalent combinations. The methylene peak is split into a quartet. The areas of the peaks in the quartet have the ratio 1 : 3 : 3 : 1.

In a *first-order* spectrum (where the chemical shift between interacting groups is larger than their coupling constant), interpretation of splitting patterns is quite straightforward;

The multiplicity of a multiplet is given by the number of equivalent protons in neighbouring atoms plus one, i.e. the n + 1 rule

Equivalent nuclei do not interact with each other. The three methyl protons in ethanol cause splitting of the neighbouring methylene protons; they do not cause splitting among themselves

The coupling constant is not dependant on the applied field. Multiplets can be easily distinguished from closely spaced chemical shift peaks.

Note 6:

The following exercises may also be tried:

A NMR spectrum of a molecule of AB type recorded in a spectrometer with frequency of 40 MHZ is given in R.S.Drago, Physical methods in Inorganic Chemistry, Reinhold, NY 1968 (Fig. 8) in page 47 (a) calculate J and $\Delta_B - \Delta_A$. (b) calculate the difference in δ for the two peaks. (c) If the spectrometer frequency is made to 100 MHZ what will be the values of J and $\Delta_B - \Delta_A$?

EXPERIMENT 100
Molecular Modeling

Aim: To practice to generate molecular models on a personal computer and using the model to arrive at the molecular geometry and other molecular parameters like energy, electron density, vibrational frequency, etc.

Equipment Required: Personal computer and relevant software such as Spartan, Winmopac or similar kind.

Molecular Modeling: Why do we need computer modeling?

Chemists have traditionally synthesized molecules and then tested their properties. In the pharmaceutical industry, for example, it was common for a company to produce hundreds of compounds for every one worth pursuing into the clinic. This is naturally very expensive and time consuming. On the other hand if a compound with the desired properties can be designed *a priori* by executing an appropriate computer program, there are obvious benefits both intellectually and commercially. This hope has become a reality now. All major pharmaceutical and agrochemical companies now employ specialists in computer-aided molecular design and the synthetic chemists are also increasingly using these facilities for themselves.

Before the advent of computers, chemists used models to aid their understanding of molecules. Basically these were of two types: wire models which represent bonds by fine metal tubes giving an idea of the molecular skeleton, and space-filling models where each atom is represented by a sphere of the appropriate radius and give an idea of the maximum of the electronic charge cloud. These types of representations have been carried over into the computer graphic molecular modelling programs. There are, however, major advantages in the computer displays over mechanical models. Molecular geometries can be drawn from crystallographic databases so that they are realistic rather than average values of bond lengths and angles; the display is created in moments rather than weeks for large molecules; the images may be manipulated interactively and parts may be removed or added. It is possible, for example, to 'observe' the active site of an enzyme from inside the protein. In fact the most sophisticated modelling programs rival flight simulators, which is not surprising since the hardware and many aspects of the software have much in common.

Most drugs and agrochemicals are small molecules of less than fifty atoms. Almost always the molecules are flexible and evoke their important biological response by binding to a specific site on a macromolecule, generally a protein or nucleic acid. Although it is possible to learn about the shape or conformation of the small binding partner from X-ray crystallography of the compound, or by nuclear magnetic resonance spectroscopy of solutions, these data are insufficient to understand the shape of molecules when they bind into the receptor slots of the macromolecule. The small molecules, and indeed the protein, may adopt their shapes on binding. Hence it becomes important to calculate the energy of the drug as a function of its shape so as to be able to answer the question into what shapes could this molecule be distorted on binding without expending more than a given energy threshold or braking the molecule? The computer performs the calculation and a graphics terminal can display the results.

Calculating Molecular Properties: Two distinct approaches are used in the computation of the energy of a molecule as a function of its shape as reflected in terms of bond distances and bond angles. One is the use of molecular quantum mechanics to solve the Schrödinger equation in an approximate Hartree-Fock manner. These computations may be of the so-called *ab initio* method in which all the many millions of integrals implied in the theory are rigorously evaluated. Alternatively a semi-empirical calculation can be made, in which time is saved by replacing computed values of integrals by experimentally determined parameters. Quantum mechanical calculations are the favoured approach if, in addition to a value of the energy or stability of a molecule, one requires some other molecular properties which can be derived from the molecular wavefunction (such as dipole moment, electrostatic potential or energies of individual molecular orbitals).

A second approach is to use the molecular mechanics methods. These are purely empirical. The methods treat a molecule as a collection of balls (the atoms) and springs (the chemical bonds). Potentials are introduced for all atom-atom interactions; for the bending of bond angles and for torsional changes. The number of disposable parameters is fixed by forcing agreement with experimental quantities, such as heats of formation. But, over the years these potentials have been refined to the point where molecular calculations are reliable for many types of molecules even when they contain heteroatoms like sulphur or rings of atoms.

Once the energy is calculated, the range of shapes which a molecule may adopt may be compared for molecules which fit a receptor site of interest with the hope of learning something about the demands of the binding process.

Structure-Activity Relationships: In addition to computing the energy of an isolated small molecule, other calculable properties may be of utility, especially in the search for a relationship between molecular structure and biological activity. From the molecular wavefunction which is, along with molecular energy, the output from a quantum mechanical calculation, a large variety of molecular properties may be calculated.

The square of the function at any point in the space surrounding a molecule is, of course, a measure of electron density. One may also compute at the same point the interaction energy of a unit charge (the electrostatic potential) or its gradient, the electrostatic field, which is a measure of the interaction energy a dipole would have at that position. Computer graphics permit representations of these important quantities to be made using colour coding and stereoscopic views. Such graphical displays permit qualitative connections to be made between properties and activity.

For quantitative structure activity relationships, the numerical values of computed quantities are used in regression analyses to provide statistically supported predictions of the activity of hypothesized new compounds. The parameters used in this way include the charges on specific atoms; the electron density in bonds; the availability of electrons as judged by orbital energies; the capacity to accept electrons and perhaps most effectively the properties of the so-called frontier orbitals - those electrons which are least tightly bound by the molecule.

Protein-Drug Binding Calculations: The use of computers has concentrated on calculating and displaying properties of the small molecules which act biologically by binding to receptor sites on macromolecules. Computational techniques have been developed to the extent that the more logical approach of including the protein or nucleic acid macromolecule is now also possible. The first requirement is knowledge of the molecular structure of the large molecule. In ideal circumstances this comes from X-ray crystallographic studies of crystals of the macromolecule. High resolution refined crystal structures are now available for large numbers of proteins, mostly enzymes and for some oligonucleotides.

Based on these crystal structures the architecture of proteins can be derived by a combination of computer techniques including recognizing similar structural elements, theoretical calculation of relative energies and computer-aided model building. This type of work is important as gene sequences (and hence the amino acid secondary structure of proteins) become more readily available.

If one has a reasonable knowledge of both - the macromolecular binding site and the small molecular partner then the calculation of their binding energy is an important quantity which can come from theoretical computation. If one wants to design an inhibitor of an enzyme then that inhibitor must bind to the active site more tightly than the natural substrate. If that inhibitor is to act as a drug then the stronger it binds, the smaller the dose which will be required medicinally. The actual calculation of binding energy can be done using the same types of computer program which are used to calculate the energetic properties of isolated small molecules.

In quantum mechanical calculations, inclusion of every atom of receptor protein is too demanding in terms of computer time, so approximations such as treating atoms as point partial charges are introduced. The molecular mechanics calculations by contrast are rapid enough to cope with the tens of thousands of enzyme atoms. Both approaches have now reached a level of sophistication where the calculations of binding energy are in good agreement with experimental binding enthalpies so that novel inhibitors can now be designed.

Introduction of molecular motion:

So far molecular design has been treated almost like engineering design: molecules must fit together and bind tightly but they have been considered to have definite static shapes. In reality, of course, molecules are in dynamic motion. When a small molecule binds to an enzyme, it is not a key fitting a lock but a throbbing vibrating pair of species which interact. These dynamic effects can no longer be ignored. Since the molecular mechanics potentials can be computed readily, the gradients of the potential and dynamic behaviour may be predicted using Newton's laws of motion. This use of so-called molecular dynamics has become particularly important now that experimental dynamic properties are emerging from nuclear magnetic resonance studies. This has given a level of realism to computer-aided molecular design and is clarified by the use of computer graphic displays to illustrate the motions involved.

Procedure: This experiment is run on a PC. *No special knowledge of computing is required.* The experiment is in two parts. In the first, one will use the *molecular modelling package Spartan* (or any other similar packages) to complete a number of straight forward tasks, whose aim is to allow one to become familiar with the software. In the second, one will devise an exercise which uses Spartan to solve some appropriate chemical problems.

The project forms the bulk of this experiment, and may require several hours work with the modelling program and possibly some time spent investigating the relevant literature.

Once a suitable exercise is chosen,

(i) use the molecular modelling package to investigate it;

(ii) prepare a suitable write-up, discussing the results, comparing them to any found in the literature and considering their reliability.

Carefully prepare a project report and also discuss if any extra features of Spartan that have been used so that they will be useful for future students.

Starting Spartan: Spartan is a powerful program which can be run on a PC or Unix workstation. It requires the presence of a hardware key (a "dongle") connected to the rear of the computer. To start Spartan, double click on the Spartan icon. The principal aim of this experiment is to use Spartan to solve a real chemical problem. In order to do this, one will need to become familiar with various aspects of the program, and the best way to do this is through the completion of tasks which illustrate the capabilities of the software. Print out or record in the data book any graphics or data which seem interesting.

Task 1. Building and manipulating a small molecule:

To get one started, one can first build and display a small molecule, for example bromo-methoxy-methane, ($CBrH_2 - O - CH_3$) or any other molecule of interest. Once the molecule is built, one will determine its equilibrium geometry and dipole moment. New molecules are built by selecting atomic fragments from a menu and joining them as required. As one will discover later, it is possible to specify geometry, stereochemistry, charge, multiplicity and other characteristics of the molecule.

(a) Click the left hand mouse button (lmb) on the *File* menu and select *New*.

(Alternatively, you can click on the *New* icon, which is the leftmost icon.) A panel of common molecular fragments will appear, with tetrahedral carbon already highlighted. Move the cursor to the middle of the green display area and click the lmb to place a tetrahedral carbon as the starting point for the structure.

(b) Click on the *Br* box in the fragment panel and then click the lmb over the end of one of the carbon single bonds to add the bromine.

(c) Select from the fragment panel the singly-bonded oxygen atom and add that to the structure.

(d) Finally add a second tetrahedral carbon to the free oxygen bond. All remaining free valencies will automatically be completed with hydrogen atoms.

You can rotate the molecule by pressing and holding down the lmb in the display area and moving the mouse. One can move the molecule using the rmb. Try both now.

(e) You can determine the molecular mechanics strain energy by selecting *Minimise* from the *Build* menu. (Alternatively, you can use the icon for this, which is the *E* with a downward-pointing arrow). The geometry will adjust rapidly as the calculation is in progress, and the final strain energy will be displayed at the bottom right, together with the molecular point group. Record both of these in ther data book. The molecule can be displayed on screen in several different formats.

(f) Choose *View* from the *Build* menu and note the effect of selecting different types of display from the *Model* menu. The molecule can be rotated and moved in each display type. Print out one of the structures if needed.

(g) Return to the ball and wire model. In this format (and in the wire model) it is possible to display atom and bond labels. Select *Configure Labels...* from the *Model* menu and make sure that *Labels* in the *Model* menu is ticked. Click *OK*, and select *Labels* from the *Model* menu. You can remove labels by selecting *Labels* from the *Model* menu.

Task 2. Determining molecular geometry:

It is simple to make measurements of the geometric parameters for the molecule.

(a) Return to the ball-and-spoke model. Select *Measure Distance* from the *Geometry* menu. Click the mouse on two atoms in succession, or (to measure the distance between bonded atoms) on the bond which joins them to display the distance between the atoms. Measure the length of each of the C – O bonds and record the distances in the data book. Comment on the values.

(b) Determine the C-Br bond length and compare this to the average value quoted in any standard chemistry text book.

(c) Choose *Measure Angle* from the *Geometry* menu. Click three atoms in succession (or on two bonds sharing a common atom) to determine the angle between the two bonds. Determine the angles around the bromine-bonded carbon atom and comment on their values.

Task 3. MO calculations:

A variety of parameters can be found through a MO calculation. The first step required is to choose the type of calculation.

(a) Select *Calculations...* from the *Setup* menu. In the dialogue box which opens up select *Equilibrium geometry* from the leftmost pull-down menu. Select *Hartree-Fock* and *3-21G* from the two rightmost menus to specify a Hartree-Fock calculation using the 3-21G split-valence basis set. Verify that the *Total Charge* is *Neutral* and the *Multiplicity* is *singlet*. Click on *OK* to remove the dialogue.

(b) Select *Submit* from the *Setup* menu. A file browser appears. Choose the default name (probably "Spartan1" or something similar) and click on *Save*. This step will start the calculation off-line. Click *OK* to remove the message.

(c) Once the calculation has completed (between ten seconds and several minutes will be required depending upon the complexity of the calculation and speed of the machine) it will be notified. One can inspect output from the job using *Output* in the *Display* menu. Of particular interest is the dipole moment. To find this, select *Properties* from the *Display* menu. Record the dipole moment and compare it to the literature value for this molecule of 2.06 Debye.

Task 4. Displaying electron density and potential energy surfaces:

(a) Select *Surfaces* from the *Setup* menu. Click on *Add...* Now select *density* from the *Surface* menu and *none* from the *Property* menu. Click a second time on *Add...* and this time select *density* from the *Surfaces* menu and *potential* from the *Property* menu. This specifies the calculation of an electron density surface onto which the value of the electrostatic potential has been mapped. Click on *OK*.

(b) Submit the job (*Submit* from the *Setup* menu); you do not need to remove the dialogue window to do this. When the calculation is complete again select *Surfaces*.

Click on the line *density completed 0.002*. Click on the density surface. A menu will appear at the bottom right of the screen. Note the effect of changing the display style and print out one of the figures.

Select *Close* from the *File* menu to remove the molecule from the screen.

Task 5. Specifying molecular conformations; calculating and displaying Molecular Orbitals:

Depending upon the project proposed, one may need to specify the conformation of a molecule, or calculate and display molecular orbitals. This section explains how those tasks can be completed.

(a) Construct, for example trans-acrolein ($CH_2 = CHCHO$) as follows: Remove any previous molecule that has been present on the screen. Select *New* from the *File* menu and click on trigonal planar sp^2 hybridised carbon. Click on the screen, rotate the molecule if necessary to identify the free double bond and click again on that to make ethylene. Click on any of the free valances to add the third carbon, then finally add a doubly-bonded oxygen.

(b) In the trans molecule the two double bonds point in the same direction. If the molecule you have made is in the cis configuration, change the conformation as follows: Click on the \?\ icon in the icon bar. Click in order on the oxygen and the three carbons. The value of the dihedral angle which these atoms define will appear in a box at the bottom right. Replace 0 by 180 and press the return key; the correct conformation will be created. Acrolein and other α, β-unsaturated carbonyl compounds are known to undergo both nucleophilic addition at the carbonyl carbon and Michael addition at the olefin position.

One can determine which is more likely in acrolein by calculating and displaying the LUMO, into which it is likely an incoming pair of electrons in a nucleophile will be directed. (The carbon on which the LUMO is most localised will be the most susceptible to attack by a nucleophile).

(a) Enter the *Surfaces* dialogue. Click on *Add...* and select *LUMO* from the *Surface* menu. Click on *OK* and submit the job.

(b) When the job has completed, click on the line *LUMO Completed* 0.032 inside the *Surfaces* dialogue.

Print out the resulting surface and comment on whether this is consistent with experimental evidence for reactivity of α, β-unsaturated carbonyl compounds.

Task 6. Calculation and display of vibrational frequencies:

It is possible to calculate and display the vibrational motions of molecules using Spartan. Let us do it for acetylene. Follow the steps.

(a) Remove any molecules from the screen, then construct acetylene (ethyne).

(b) Select *Calculations...* from the *Setup* menu. Select *Equilibrium geometry* from the leftmost menu and *Hartree-Fock* and *3-21G* from the two rightmost menus. Click on the *Frequencies* box, then click on *OK* and submit the job.

(c) Once the job has completed, display the vibrational frequencies by selecting *Vibrations* under the *Display* menu. Make a note of the predicted vibrational frequencies with their symmetry, and compare with those measured experimentally for acetylene (in the low resolution spectrum of acetylene, bands appear centred around (750, 1250, 1630 and 3200 cm^{-1}). Calculated frequencies are typically high by around 10 % and it is common practice to scale frequencies (by 0.9 in the case of the 3-21G basis set) to bring them more into line with experimental data. One should also bear in mind that combination and difference bands, though not calculated by this program, may appear in the spectrum.

(d) To animate a vibration, click on its frequency in the vibrations dialogue. One can stop the animation by clicking a second time on the frequency. One may adjust the display style to give the clearest motion.

Check that the form of each vibration is consistent with what one would expect from the symmetry shown in the vibrations table. Sketch the form of each vibration, indicate which frequency is associated with which vibration, and identify those which should be Infrared active.

Task 7. Building more complex molecules:

Spartan contains a small library of pre-formed molecules and fragments, which simplify the building of complex molecules. A few examples will illustrate how one can construct large molecules.

(a) *Pydrine*: Click on *New* in the *File* menu. Click on *Rings* in the model kit and select *benzene*. Place it on the screen. Select aromatic nitrogen (nitrogen with two single bonds and a dotted third bond) from the atomic fragments, position the cursor over a ring carbon and double click to make pyridine.

(b) *Furufral:* Click on *Rings* in the model kit and select a cyclopentane ring. Choose an Oxygen atom with two single bonds and double click on a ring carbon. To make the double bonds, select *Make Bond* from the *Build* menu. Click on one of the free valences of each of the two carbons you wish to join to make the double bond.

Click on the minimise energy icon, then add the carbonyl group from the *Groups* menu.

(c) *Camphor:* Remove any molecules from the screen and click on *New* in the *File* menu. Place a cyclohexane ring on the screen. Place an sp^2 hybridised carbon in the ring, then add the carbonyl oxygen. Click on sp^3 hybridised carbon and click on an axial free valance on one of the carbons adjacent to the carbonyl group. Click on the make-a-bond icon (8th from the left). Click on one of the free valances of the atom you have just added and then on the equatorial free valence on the carbon on the opposite side of the ring. Create a refined structure by clicking on the minimise icon. Then add three methyl groups to complete the structure.

(d) *Sulfur tetrafluoride*: SF_4 is a simple molecule to construct, but cannot be done from the standard model kit since sulfur is not in its normal bent dicoordinate geometry, but in a trigonal bipyramidal form. Bring up the expert model kit by clicking on the expert tab at the top of the model kit. Click on *S* in the periodic table, check in the window above the table that it is 5-coordinate; if it is not, select trigonal pyramidal co-ordination from the buttons below the table. Place the five co-ordinate sulfur anywhere on screen. Select *F* from the table and the one-coordinate entry-J from the list of atomic hybrids. Click on both of the axial free valences of sulfur and two of the equatorial free valences. You need to delete the remaining free valance on sulfur, otherwise it will be assumed to be attached to a hydrogen atom. Click on the delete symbol (the 7 th icon from the left) and then on the remaining free valence. Run the molecular mechanics minimisation, and check that the molecule has *C2v* symmetry. If the symmetry is T_d instead, exit the program and restart.

Other facilities:

Spartan also allows the calculation of

1. low-energy conformations using Monte Carlo methods,
2. enthalpies of reaction,
3. barriers of rotation,

4. electron distribution in LUMOs and HOMOs,
5. properties of intermolecular complexes,
6. the location of transition states,
7. the variation of energy as reaction occurs, and
8. the properties of molecules with unpaired electrons.

Details of how to accomplish these tasks may be obtained from the Spartan Pro tutorial and one can familiarize himself.

APPENDIX I

The System of SI Units

Catagory	Name	Abbreviation
Length	meter	m
Mass	kilogram	kg
Time	second	s
Electric current	ampere	A
Temperature	kelvin	K
Amount of substance	mole	mole
Luminous intensity	candela	cd

APPENDIX II

SI Units and Conversion Factors

To convert from	To	Multiply by
	Acceleration	
Feet per second per second	centimeter per second per sec	30.480
	meter per second per second	0.30480
Kilometer per hour per second	centimeter per second per sec	27.778
	feet per second per second	0.9113
	meter per second square	0.27778
	Density	
Gram per cubic centimeter	(Mass/Volume)	
	kilogram per cubic meter	0.001

To convert from	To	Multiply by
	Energy and work	
British thermal units	joules	1055
British thermal units per second	watts	1.055
British thermal units per minute	horsepower	0.02358
	watts	17.58
Calories	British thermal units	0.0039683
	foot-pounds	3.088
	joules	4.1868
	kilogram-meters	0.4265
	watt-hours	0.001163
Ergs	joules	0.0000001
Joules	British thermal units	0.0009484
	calories	0.2390
	foot-pounds-force	0.73756
	watt-hours	0.00027778
Kilowatts	British thermal units per min.	56.92
	foot-pounds-force per minute	44254
	foot-pounds-force per sec.	737.6
	horsepower	1.3410
	kilocalories per minute	14.34
	Force	
Dyne	newtons	0.00001
Kilograms-force	newtons	9.8066
	pounds-force	2.2046
	Length	
Centimeters	inches	0.3937
Feet	centimeter	30.480
	inches	12
	meters	0.30480
Inches	centimeter	2.540
	meters	0.02540
	millimeters	25.40
Kilometes	feet	3280.8
	miles	0.62137
Micrometers	meters	0.000001
Millimeters	inches	0.039370
Miles	feet	5280
	kilometers	1.6093
	meters	1609.3
	yards	1760

To convert from	To	Multiply by
	Pressure and Stress	
Atmosphere	bars	1.01325
	centimeters of mercury at 32 F	76.0
	feet of water at 68 F	33.96
	inches of mercury at 32 F	29.921
	kilograms-force per sq. cm^2	1.0332
	kilograms-force per sq. m.	103322
	kilopascals	101.325
	pounds-force per sq. inch	14.696
	tons-force (short) per. sq. foot	1.0581
	torr	760
Centimeters of mercury	atmospheres	0.013158
	bars	0.01333
	feet of water at 68 F	0.4468
	inches of water at 68 F	5.362
	kilograms-force per sq. cm^2	0.19337
	pound-force per sq. inch	27.85
	torr	10
Kilograms-force per sq. cen.	atmospheres	0.9678
	bars	0.98066
	feet of water at 68 F	32.87
	inches of mercury at 0° C	28.96
	kilopascals	98.066
	pounds-force per. sq. foot	2048
	pounds-force per. sq. inch	14.223
Kilopascals	dynes per sq. centimeter	10000
	feet of water at 68 F	0.3351
	inches of mercury at 32 F	0.2953
	inches of water at 68 F	4.021
	kilograms-force per. sq. cm^2	0.010197
	pascals	1000
	pounds-force per. sq. inch	0.1450
Millibars	pascals	100
Millimeters of mercury at 0°C	bars	0.0013332
	feet of water at 68 F	0.004680
	inches of mercury	0.03937
	inches of water at 68 F	0.53616
	kilograms per sq. centimeter	0.0013595
	pascals	133.32
	pounds per sq. inch	0.0193368

To convert from	To	Multiply by
	Velocity	
Centimeters per second	feet per second	0.03281
	feet per minute	1.9685
	miles per hour	0.02237
	kilograms per hour	0.03600
	meters per minute	0.60000
Kilometers per hour	centimeters per second	27.778
	feet per second	0.9113
	feet per minute	54.68
	international knots	0.53996
	meters per minute	16.667
	meters per second	0.27778
	miles per hour	0.6214
Kilometes per second	miles per minute	37.28
Meters per second	feet per minute	196.8
	feet per second	3.281
	kilometers per hour	3.600
	kilometers per minute	0.0600
	miles per hour	2.237
	miles per minute	0.03728
Miles per hour	centimeters per second	44.70
	feet per minute	88
	feet per second	1.4667
	international knots	0.8690
	kilometers per hour	1.6093
	meters per minute	26.82
	Volume	
Cubic centimeters	cubic inches	0.06102
	cubic feet	0.000035315
	cubic yards	0.000001308
	gallons (US)	0.0002642
	gallons (imp)	0.00022
	liters	0.001
Cubic feet	cubic centimeters	28317
	cubic meters	0.028317
	cubic inches	1728
	cubic yards	0.03704
	gallons (US)	7.4805
	gallons (imp)	6.229
	liters	28.32
To convert from	To	Multiply by

To convert from	To	Multiply by
Cubic meters	cubic inches	61024
	cubic feet	35.315
	cubic yards	1.3080
	gallons (US)	264.17
	gallson (imp)	219.97
	liters	1000
Gallons (US)	cubic centimeters	3785.4
	cubic meters	0.0037854
	cubic inches	231
	cubic feet	0.13368
	cubic yards	0.0049515
	pints (liquid)	8
	quarts (liquid)	4
	gallons (imp)	0.8327
	liters	3.7854
	pounds of water at 60 F	8.338

APPENDIX III

Concentration of Acids

Acids	Specific gravity	Percent by weight	Approximate Normality
Conc. HCl	1.19	37.9	11
Conc. H_2SO_4	1.84	96.01	36
Conc. HNO_3	1.41	69.8	16
Galacial CH_3COOH	1.06	99.51	17

APPENDIX IV

Physical Constants of Some Solvents at 20°C

Solvent	Molecular weight	Refractive index (sodium D line)	Density (g/cm^3)
Acetic acid	60.05	1.372	1.040
Aniline	93.12	1.586	1.622
Acetone	58.08	1.359	0.796
Benzene	78.11	1.501	0.879
Carbon tetra chloride	153.84	1.463 (15°)	1.595
Chloroform	119.39	1.446 (180°)	1.498 (15°)
Ethyl alchol	46.07	1.361	0.785 (25°)
n-Heptane	100.20	1.388	0.684
n-Hexane	86.19	1.375	0.660
Methyl alcohol	32.04	1.329	0.792
Toluene	92.13	1.497	0867

APPENDIX V

Dissociation Constants of Weak Acids and Weak Bases in Water at 25°C

Acetic acid	CH_3COOH	1.74×10^{-5}
Benzoic acid	C_6H_5COOH	6.3×10^{-5}
Chloroacitic acid	$ClCH_2COOH$	1.35×10^{-3}
Dichlorocetic acid	$Cl_2CHCOOH$	5.0×10^{-2}
Formic acid	$HCOOH$	1.76×10^{-4}
Oxalic acid	$H_2C_2O_4$	6.5×10^{-2}, 6.1×10^{-5}
Phenol	C_6H_5OH	1.6×10^{-10}
Ammonia (aq)	NH_4OH	1.74×10^{-5}
Aniline	$C_6H_5NH_2$	4.30×10^{-10}
Ethylamine	$C_2H_5NH_2$	4.4×10^{-4}
Hydroxilamine	NH_2OH	9.1×10^{-9}
Methylamine	CH_3NH_2	4.2×10^{-4}
Pyridine	C_5H_5N	2.0×10^{-9}

APPENDIX VI

Solubility Product of Some Sparingly Soluble Salts at 25 °C

Substance	Solubility product
Barium carbonate	8.1×10^{-9}
Barium sulphate	1.08×10^{-10}
Mercurous chloride	2.0×10^{-18}
Silver chloride	1.5×10^{-10}
Silver chromate	9.0×10^{-12}

APPENDIX VII

Ionic Conductance at Infinite Dilution (λ_0) of Ions at 25 °C

Ion	Ag^+	Cl^-	H^+	K^+	Na^+	NH_4^+	OH^-
λ_0	63.5	75.5	350	74.5	50.9	74.5	192

APPENDIX VIII

Specific Conductivity of KCl Solutions

Temperature (°C)	N/10 KCl	N/100 KCl
20	0.01167	0.09128
22	0.01215	0.001332
24	0.01264	0.001386
26	0.01313	0.002819
28	0.01362	0.002927
30	0.01412	0.003036

APPENDIX IX

Surface Tension of Water

Temperature (°C)	5	10	20	30	40	50	60	70	80
Surface Tension (dyne / cm)	74.9	74.2	72.8	71.2	69.6	67.9	66.2	64.4	62.6

APPENDIX X

Surface Tension (in dyne/cm) of Some Common Solvents

Substance	Surface tension at 20°C	Surface tension at 50°C
Acetic acid	27.6	24.7
Acetone	23.7	—
Benzene	28.9	25.6
Carbon tetrachloride	26.8	23.1
Chlorobenzene	33.2	29.6
n-Hexane	22.6	20.1
Ethyl alcohol	22.3	19.8
Methyl alcohol	28.4	25.0

APPENDIX XI

Viscosity of Some Common Liquids at 25 °C

Liquids	Viscosity (centipoise)
Acetic acid	1.155
Acetone	0.316
Aniline	3.71
Benzaldehyde	1.39
Chloroform	0.542
Ethyl acetate	0.441
Glycerol	954
n-Heptane	0.386
n-Hexane	0.294
Methanol	0.547

APPENDIX XII

Viscosity of Water at Different Temperature

Temperature (°C)	5	10	15	20	25	30	35
Viscosity (cp)	1.5188	1.3077	1.1404	1.0050	0.8937	0.8007	0.7225

APPENDIX XIII

Buffer Solutions

(a) Acetic Acid — Sodium Acetate Buffer

Stock solutions: 0.2 M acetic acid and 0.2 M sodium acetate

pH	3.5	4.0	4.5	5.0	5.2	5.5	5.6
Acetic acid (mL)	94	82	56	30	20	12	10
Sodium acetate (mL)	6	18	44	70	80	88	90

(b) Phosphate Buffer

Stock solutions: M/15 potassium dihydrogen phosphate and M/15 disodium hydrogen phosphate

pH	5.5	6.0	6.5	7.0	7.5	8.0
KH_2PO_4 (mL)	96	88	69	38	15	6
Na_2HPO_4 (mL)	4	12	31	62	85	94

(c) Borate Buffer

Stock solutions: 0.05 M borax and 0.2 M boric acid

pH	7.0	8.0	9.0	10	11
Borax (mL)	6.0	27.5	80.0	97.5	98.4
Boric acid (mL)	94.0	72.5	20.0	2.5	1.6

APPENDIX XIV

Electromagnetic Spectrum

Electromagnetic radiation such as visible light is commonly treated as a wave phenomenon, characterized by a wavelength or frequency. Wavelength is defined as the distance between adjacent peaks (or troughs), and may be designated in meters, centimeters or nanometers (10^{-9} meters). Frequency is the number of wave cycles that travel past a fixed point per unit of time, and is usually given in cycles per second, or hertz (Hz).

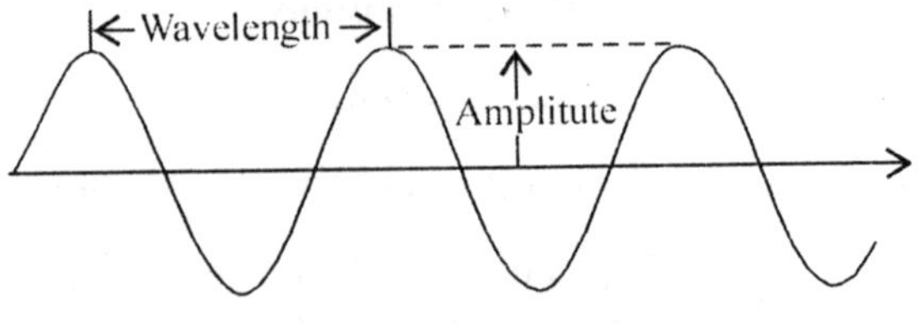